Capture Dynamics and Chaotic Motions in Celestial Mechanics

Capture Dynamics and Chaotic Motions in Celestial Mechanics

With Applications to the Construction of Low Energy Transfers

Edward Belbruno

PRINCETON UNIVERSITY PRESS
PRINCETON AND OXFORD

Published by Princeton University Press, 41 William Street,
Princeton, New Jersey 08540

In the United Kingdom: Princeton University Press, 3 Market Place,
Woodstock, Oxfordshire OX20 1SY

The publisher would like to acknowledge the author of this volume
for providing the camera-ready copy from which this book was printed.

Library of Congress Cataloging-in-Publication Data

Belbruno, Edward, 1951-
Capture dynamics and chaotic motions in celestial mechanics:
with applications to the construction of low energy transfers / Edward Belbruno
p. cm
Includes bibliographical references and index
ISBN 0-691-09480-2 (cl : alk. paper)
1. Many-body problem. 2. Chaotic behavior in systems.
3. Celestial mechanics. I. Title.

QB362.M3B45 2004
521-dc21 2003056475

British Library Cataloging-in-Publication Data is available

Printed on acid-free paper. ∞

www.pupress.princeton.edu

10 9 8 7 6 5 4 3 2 1

Contents

List of Figures

Foreword

One of the most interesting things to happen in astrodynamics recently is the resurgence of interest by people in the space sciences communities in the deeper mathematical aspects of the N-body problem. There are many reasons for this, including the fact that many current space missions have truly complex and highly non-Keplerian orbits that involve the subtleties of the *three-* and N-body problems. Some of these missions are *The International Sun-Earth Explorer 3* (ISEE-3) [aka *International Cometary Explorer* (ICE)], the *Solar and Heliospheric Observatory* (SOHO) Mission, the *Near Earth Asteroid Rendezvous* (NEAR), the *Genesis Discovery Mission*, and the *Wilkinson Microwave Anisotropy Probe* (WMAP), to name a few. Moreover, many of the missions planned for the future (the *Terrestrial Planet Finder* (TPF) and the *Jupiter Icy Moons Orbiter* (JIMO), to name just two) will also make use of interesting orbit structures. All of these missions can be read about in more detail simply by "googling" their names.

For those of us who have an interest in things both celestial and dynamical, this is very gratifying and is a fitting tribute to one of the founders of it all, namely Poincaré. In fact, some future mission could rightly be named after him for his genius and foresight. After all, it was Poincaré who, motivated by his struggles with the nonintegrability of the three-body problem around 1890, created the modern era of dynamical systems, including the idea of chaotic trajectories, heteroclinic orbits, and separatrix splitting. Of course, Poincaré himself built on the work of Euler (who discovered the collinear libration points L_1, L_2, L_3 and Lagrange (who discovered the triangular points L_4 and L_5) and many others.

It is worth recalling that it was the problem of stability of the triangular Lagrange points L_4 and L_5 that motivated the study of KAM (Kolmogorov-Arnold-Moser) theory. This theory has undergone many profound mathematical developments, such as the emergence of Aubrey-Mather theory that the reader will encounter in chapter 2 of the present book. These mathematical developments have had an enormous impact on both dynamical systems and celestial mechanics. More generally, the chaotic nature of solutions to the N-body problem has coupled with many deep insights into the workings of the solar system, such as Wisdom's work on the Kirkwood gaps. Mathematics, especially the idea of symplectic transformations, also helped to provide some of the key ingredients in the construction of the long time dynamics of the solar system and planets by people such as Wisdom, Sussman, Laskar, Tremaine, Quinn, and Duncan.

It is also quite interesting how specific missions have motivated the fundamental mathematical theory and how the mathematical theory in turn is intertwined with the mission. An example is how the ICE mission was closely connected with Farquhar's original studies of halo orbits. Another excellent example of this concerns the "rescue" of the *Hiten* (originally called *MUSES-A*) mission, which was intentionally crashed into the Moon on 10 April 1993. The "rescue" was achieved by Belbruno and Miller by making use of four–body dynamics, which led to Ed Belbruno's ideas involving the WSB (weak stability boundary), which the reader will find discussed in chapter 3 of Ed's book. Another example of this sort was the spectacular *NEAR* mission, headed by Robert Farquhar, which landed a spacecraft on the asteroid EROS. This sort of mission gives much information and inspiration to the study of asteroid and Kuiper belt binaries, which is now a subject of great interest to astrodynamicists and is one that geometric mechanics can contribute to in a significant way.

Another, more personal, example of the interaction of mathematics and mission design was how the *Genesis Discovery Mission* together with the *Hiten* work of Belbruno and Miller and the work of Belbruno and Brian Marsden on resonance in comets, led to some of my own work with Martin Lo, Wang-Sang Koon, and Shane Ross. We endeavored to put the old (1963) invariant manifold tube ideas of Conley and McGehee into the context of astrodynamics and mission design and to tie all these ingredients together. This started out as primarily a mathematical exercise in understanding how all the pieces of the jigsaw puzzle fit together to form a coherent picture. Our ideas eventually led to some concrete new ideas and proposals, such as our idea for a *Multi-Moon Orbiter*, published earlier this year in the proceedings of an American Institute of Aeronautics and Astronautics conference. We hope that the *JIMO* mission will come to pass someday and that it will further show how mathematics and concrete missions are cooperative and helpful players in the common goal of better understanding as well as exploring our solar system.

It is a great pleasure to write these few brief words of introduction to Ed's book. I hope that you, the reader, will enjoy reading the book and find it helpful and inspiring.

Jerry Marsden
Pasadena, California
Spring 2003

Preface

Introductory Remarks

Celestial mechanics is an old field and has its roots in the accurate mathematical description of the motions of the planets. This goes back, in particular, to the important work of the Greek astronomer Ptolemy in 150 A.D. who used the concept of epicycles to describe the motions of the Moon and planets known at that time [178]. His method was to superimpose a sufficient number of epicycles, which are circular motions with different periods, in order to accurately describe a given planetary motion. In contemporary terminology, this is equivalent to uniformly approximating almost periodic functions by superposition of quasi-periodic functions, which is validated by a theorem of Bohr [213, 47]. With Newton's formulation of the gravitational inverse square force law [179], together with the calculus and Kepler's work [125], in the 17th century a more systematic mathematical approach to describing the motions of the planets became available. Celestial mechanics took on a more mathematical form after the time of Newton, with the work of Euler [80] and Lagrange [132] in the 18th and 19th centuries, respectively. Poincaré's fundamental mathematical work at the end of the 19th century provided revolutionary new methods for studying celestial mechanics using what is known today as dynamical systems theory [190]. A basic model he used to apply his methods is the restricted three-body problem, which has remained to this time as a basic model to use, and which is still far from completely understood. This model is a main focus of this book.

A central theme of this book in general is to study different types of chaotic and complicated dynamics in the three-body problem and relate to them some new results.

This book has several goals. One goal is to provide a way to unify two seemingly different types of capture processes and associated dynamics. Capture in the three-body problem has played a fundamental role in the field of celestial mechanics, where one mass particle is somehow bound to one or both of the other mass particles. The definitions of how this can happen are substantially different. One of these capture processes is called *permanent capture*, and it has played an important theoretical role in celestial mechanics by the proof that it is chaotic in nature. This proof was accomplished by Alekseev in the 1960s [6], and later by Moser [175] for a version of the three-body problem called the Sitnikov problem.

Another type of capture, called *ballistic capture*, is defined in a completely different way. It is used in a more applied setting in astronomy and astrodynamics, where it has proved to be important in recent years in finding new types of low energy transfers from the Earth to the Moon, one of which was used in practice in 1991. Permanent and ballistic capture have been viewed as being unrelated. Permanent capture is geometrically defined in a global manner and is related to unbounded motion whereas ballistic capture is analytically defined for bounded motion.

We prove that these two forms of capture are equivalent on a special set called a weak stability boundary on which exists a complicated set called a hyperbolic network. This network gives rise to chaotic dynamics in the restricted three-body problem of which permanent capture is one type of dynamics. This network has been proven to exist by Xia [227] in a different setting. The proof of equivalence of these two types of capture processes is a new result. It sheds light on the ballistic capture process. Implications and applications are discussed. The connection of ballistic capture with chaotic motion helps explain the complexity of the lunar transfer described in section 3.4, or certain types of cometary motion. Another goal of this book is to give a systematic presentation of different kinds of capture, as well as a sense of how they are related. The subject of capture is presented in chapter 3.

Another type of complicated motion we study in this book occurs near bounded motion. This bounded motion is associated with a special family of periodic orbits initially proven to exist by Poincaré called periodic orbits of the first kind. They have played a central role in the field of celestial mechanics in illustrating the complicated dynamics that occurs near them. This is because special types of two-dimensional maps can be defined in a neighborhood of orbits of the family. These are called twist maps. Using Kolmogorov-Arnold-Moser(KAM) theory, the Moser twist theorem can be applied near these orbits, yielding a sequence of invariant closed curves, which bound resonance gaps where very complicated chaotic motions occur. The Birkhoff fixed point theorem can be applied in these gaps to produce infinitely many periodic points. In this book, following in this tradition, we prove that Aubrey-Mather sets also exist in these particular gaps. We present this material in Chapter 2. Aubrey-Mather sets give rise to a very complicated motion.

Finally, another goal of this book is to present the two-body problem from the perspective of geodesic flows on spaces of constant curvature. This is done in chapter 1, where a new proof is given.

Contents

Chapter 1 provides an introduction to celestial mechanics in which the n-body problem is defined and the conservation laws are derived. The two-body problem is completely solved and collision is regularized. The restricted three-body problem is derived. Geodesic equivalent flows for the two-body problem are described and a new proof of this equivalence is presented which is substantially shorter than previous proofs. Collision and noncollision singularities are discussed for perspective.

Chapter 2 focuses on bounded motion and area-preserving twist maps. Quasi-periodic motion in the restricted three-body problem is proven to exist by the KAM theorem. The reduction to a map is next considered with a presentation of the Moser twist theorem. Area–preserving maps with fixed points are categorized. Transverse homoclinic and heteroclinic points for maps are defined. The intricate structure of an area-preserving map near an elliptic point is discussed. Of particular interest are the regions of resonance, called resonance gaps. Periodic orbits of the first kind are next proven to exist by the Poincaré continuation method, and from this an area-preserving monotone twist map is obtained. The Moser twist theorem and Birkhoff fixed point theorem are applied. We conclude the chapter with an application of Aubrey-Mather theory which is applied near the periodic orbits of the first kind.

In chapter 3 the subject of capture is considered. Different forms of capture are defined, some here for the first time. For the sake of applications, the *capture problem* is defined. The weak stability boundary is defined numerically and also analytically. Conley's work on transit orbits is described and extended. Two astrodynamics applications are discussed illustrating unstable ballistic capture, the main one being the use of a transfer to the Moon to resurrect a Japanese lunar mission. Moser's proof of chaos associated with permanent capture in the Sitnikov problem is summarized, and Xia's related proof for the restricted three-body problem is outlined. We conclude the chapter with a proof that ballistic capture on an extended version of the weak stability boundary is a chaotic process by proving the existence of a hyperbolic invariant set.

There are several new theorems and results in this book that have not been previously published. These include Theorem 2.33 on the existence of Aubrey-Mather sets in the restricted problem, and Theorem 3.58 on the proof of existence of an invariant hyperbolic network on an extended version of the weak stability boundary. A short proof is given for Theorem 1.39 on the existence of geodesic flows for the Kepler problem.

Scope of This Book

The approach taken here is mathematical and is both theoretical and applied, with an emphasis on theory. The general mathematical approach is that of dynamical systems. Theorems are stated and proven, and to illustrate some of them, applied examples are used. These examples are from astronomy and astrodynamics. The examples are more expository in nature and meant to be only illustrative. The two main examples discussed at length are from astrodynamics, and a few shorter ones are from astronomy. The strictly applied parts of this book pertaining to astrodynamics only occur in the following places in chapter 3: subsections 3.1.5 and 3.3.2, and section 3.4. The remainder of the book is presented in a formal mathematical fashion.

The contents are self-contained, and some theorems are stated without proof. In these cases the proofs are well established in the literature and are not necessary for our presentation. Proofs that are well established in the literature that are needed for our presentation are summarized for their key points. For example, Moser's proof in [175] is summarized in subsection 3.6.4. This proof by Moser is itself an entire book, and due to space limitations, we summarized it in as brief a fashion as possible in a few pages. Likewise, the paper [227] is summarized in subsection 3.6.5 so as to yield the key parts necessary for our presentation. This proof closely follows the methodology of the proof in subsection 3.6.4, and we will make comparisons to it. Theorems and new results that are formulated within the book are proved in a rigorous fashion.

Each chapter begins by describing the contents of the chapter and explaining the plan to be followed.

This book can serve as a basic textbook for celestial mechanics. Chapter 1 could serve as an introduction to the subject in an upper level undergraduate course or beginning graduate course. Chapter 2 could be included if the course were designed to be at the graduate level. Chapter 3 is approximately one half of this book and most of it, especially section 3.6, would be suitable at the graduate level. This book can also serve as a text for topics in dynamical systems. Parts of chapter 3 are also relevant for people interested in a more mathematical approach to capture relevant for astrodynamics, astronomy, and physics. Because of new results, this book also serves as a research text.

The prerequisites of the various chapters and sections vary. For chapter 1, sections 1.1–1.5, I would recommend a course in mechanics, elementary differential equations, and advanced calculus or mathematical analysis. Section 1.6 requires in addition a course in elementary differential geometry, such as that in [180]. Chapter 2 requires graduate courses in differential equations, mathematical analysis, and dynamical systems. Chapter 3 requires the same prerequisites as Chapter 2.

I have used some basic texts for reference throughout the writing of this book. In parts of chapter 1, an excellent reference is the book by Pollard [191]. Another is the foundational book by Siegel and Moser [204]. For the Levi-Civita regularization and related topics, the book by Stiefel and Scheifele is a good source [214]. Arnold's classical book [15] is a valuable reference in general. Chapter 2 refers to a large extent to [204] and [175] and for general dynamical systems, to the well–known text of Guckenheimer and Holmes [89]. In section 2.5 references include the basic paper of Mather [145] and the book by Katok and Hasselblatt [120]. Chapter 3 has a bit of new work. Section 3.6 outlines two proofs and makes considerable reference to [175] and [227]. These are just a few of the references; citations are made throughout the text. The bibliography includes works cited in the text as well as other references not cited, which are included since they are relevant.

Acknowledgments

There are several people and organizations I would like to thank for their support and interest in my work over many years which eventually enabled me to write this book.

I would like to acknowledge the former NASA Administrator, Mr. Daniel Goldin, Senator Thomas Harkin, Dr. Gerold Soffen (in memory), my graduate advisor Prof. Jürgen Moser (in memory), Prof. John Mather, Prof. Richard McGehee, Prof. Jaume Llibre, Dr. Edward Weiler, Dr. Harley Thronson, Dr. Niel de Grasse Tyson, Col. Douglas Kirkpatrick, Col. Michael D'Lorenzo, Dr. Gil Moore, Prof. Donald Saari, Dr. Herbert Shaw, Ms. Jane Barrash, Mr. Irah Donner, Mr. Howard Marks, Dr. Michael Schulhof, Dr. John Remo, Dr. Claudio Maccone, Dr. Louis Friedman, Dr. Thomas McDonough, Mr. Rex Ridenoure, Mr. Keith Gottschalk, Mr. Eric Frydler, Mr. Richard Rose, and at Princeton University Press, the editors Ms. Vickie Kearn, Ms. Alison Kalett, Ms. Beth Gallagher, Ms. Ellen Foos, and illustrator Mr. Dimitri Karetnikov. I would like to thank the reviewers for their very helpful comments, and I also would like to thank the typist Ms. Phebe Tarassov.

I would like to thank the following organizations: First, I am grateful for the interest shown in my work by the National Aeronautics and Space Administration. I would also like to thank The United States Air Force Academy, European Space Agency, the former McDonnell Douglas Corporation, the former Geometry Center of the University of Minnesota, Princeton University, University of Rome, and the Boeing Corporation.

Capture Dynamics and Chaotic Motions in Celestial Mechanics

Chapter One

Introduction to the N-Body Problem

The basic differential equations are defined that we will use throughout this book. These include the Newtonian n-body problem in section 1.1, and the planar three-body problem using Jacobi coordinates in section 1.2. In section 1.3, we derive the classical solutions for the two-body problem. In section 1.4 regularization is defined and collision is regularized via the classical Levi-Civita transformation and the Kustaanheimo-Stiefel transformation. Section 1.5 introduces the equations of motion for the restricted three-body problem in different variations and coordinate systems. This problem is the main focus of subsequent chapters. Also, we discuss briefly in section 1.1 the global behavior of solutions in the n-body problem having collision and noncollision singularities. Key results are stated, including Sundman's basic theorems and the Painlevé conjecture proven by Xia. This material serves as background introductory material and provides an historical perspective. The integrals of motion for the n-body problem are also derived. In section 1.6 geodesic equivalent flows on spaces of constant curvature are derived using the Euler-Lagrange differential equations, and their equivalence with the flow of the two-body problem is described. A new proof is given for this equivalence which is substantially shorter than previous proofs. The geodesic flows give rise to n-dimensional regularizations.

1.1 THE N-BODY PROBLEM

We consider the n-body problem, $n \geq 2$. Of particular interest will be the case $n = 3$ for the Newtonian three-body problem. Before this problem and variations of it are defined, we define the general n-body problem and discuss existence of solutions. The basic conservation laws are derived.

It is defined by the motion of $n \geq 2$ mass particles $\mathrm{P_k}$ of masses $m_k > 0, k = 1, 2, \dots, n$, moving in three-dimensional space x_1, x_2, x_3 under the classical Newtonian inverse square gravitational force law. We assume the

Cartesian coordinates of the kth particle are given by the real vector $\mathbf{x}_k = (x_{k1}, x_{k2}, x_{k3}) \in \mathbf{R}^3$. The differential equations defining the motion of the particles are given by

$$m_k \ddot{\mathbf{x}}_k = \sum_{\substack{j=1 \\ j \neq k}}^{n} \frac{G m_j m_k}{r_{jk}^2} \frac{\mathbf{x}_j - \mathbf{x}_k}{r_{jk}}, \tag{1.1}$$

$k = 1, 2, \ldots, n$, where $r_{jk} = |\mathbf{x}_j - \mathbf{x}_k| = \sqrt{\sum_{i=1}^{3}(x_{ji} - x_{ki})^2}$ is the Euclidean distance between the kth and jth particles, and $\dot{} \equiv \frac{d}{dt}$. Equation (1.1) expresses the fact that the acceleration of the kth particle P_k is due to the sum of the forces of the $n-1$ particles $P_i, i = 1, \ldots, n, i \neq k$. The time variable $t \in \mathbf{R}^1$. Equation (1.1) represents $3n$ second order differential equations. This equation can be put into a simpler form by first dividing both sides through by m_k, and expressing it as a first order system,

$$\dot{\mathbf{x}}_k = \mathbf{v}_k, \qquad \dot{\mathbf{v}}_k = m_k^{-1} \frac{\partial U}{\partial \mathbf{x}_k}, \tag{1.2}$$

where $\mathbf{v}_k = (v_{k1}, v_{k2}, v_{k3}) = (\dot{x}_{k1}, \dot{x}_{k2}, \dot{x}_{k3}) \in \mathbf{R}^3$ are the velocity vectors of the kth particle,

$$U = \sum_{\substack{j=1 \\ j \neq k}}^{n} \frac{G m_j m_k}{r_{jk}},$$

$U = U(\mathbf{x}_1, \ldots, \mathbf{x}_n)$ is a real-valued function of $3n$ variables $x_{kj}, j = 1, 2, 3$, and

$$\frac{\partial U}{\partial \mathbf{x}_k} \equiv \left(\frac{\partial U}{\partial x_{k1}}, \frac{\partial U}{\partial x_{k2}}, \frac{\partial U}{\partial x_{k3}} \right),$$

$k = 1, 2, \ldots, n$. Equation (1.2) represents a system of $6n$ first order differential equations for the $6n$ variables $x_{k\ell}, v_{k\ell}, k = 1, 2, \ldots, n; \ell = 1, 2, 3$. U is the *potential energy*. G is the universal gravitational constant.

If we assume $r_{jk} > 0$, then U is a well-defined function and is a *smooth function* in the $3n$ variables x_{jk}, where *smooth* means that U has continuous partial derivatives of all orders in the variables x_{jk} and is real analytic. For notation, we set $\mathbf{x} = (\mathbf{x}_1, \mathbf{x}_2, \ldots, \mathbf{x}_n)$. Then $\mathbf{x} \in \mathbf{R}^{3n}$. Similarly, $\mathbf{v} = (\mathbf{v}_1, \ldots, \mathbf{v}_n) \in \mathbf{R}^{3n}$. With this notation, $U = U(\mathbf{x})$.

System (1.2) is of the form

$$\dot{\mathbf{y}} = \mathbf{f}(\mathbf{y}), \tag{1.3}$$

where $\mathbf{y} = (\mathbf{x}, \mathbf{v}) \in \mathbf{R}^{6n}$, and also where $\mathbf{f} = (\mathbf{v}, m_1^{-1}\partial U/\partial \mathbf{x}_1, \ldots, m_n^{-1}\partial U/\partial \mathbf{x}_n) \in \mathbf{R}^{6n}$. Thus, the standard existence and uniqueness theorems of ordinary differential equations can be applied to (1.3), and hence (1.2).

Since $\mathbf{f} = (f_1, \ldots, f_{6n})$ is a smooth vector function of $\mathbf{y}$, then these theorems guarantee that through any initial point $\mathbf{y}(t_0) = \mathbf{y}_0$ at initial time

t_0 there exists a locally unique solution for $|t - t_0| < \delta$, where δ is sufficiently small. This can be made more precise: If the real functions f_k satisfy $|f_k| < M, k = 1, 2, \ldots, 6n$, in a domain $|\mathbf{y} - \mathbf{y}_0| < p$, then

$$\delta = \frac{p}{(1+6n)M}$$

(see [204]).

A system of integrals exist for (1.1) which can be used to reduce the dimension of the $(6n+1)$-dimensional coordinate space $(\mathbf{x}, \mathbf{y}, t)$. An *integral* is a real-valued function of the $6n+1$ variables x_{kj}, v_{kj}, t which is constant when evaluated along a solution of (1.1). Let $\mathbf{x}(t), \mathbf{v}(t)$ represent a solution of (1.1).

Definition 1.1 *A integral of (1.1) is a real-valued function* $I(\mathbf{x}, \mathbf{v}, t)$ *such that*

$$\frac{d}{dt} I\left(\mathbf{x}(t), \mathbf{v}(t), t\right) = 0,$$

where the solution $\mathbf{x}(t), \mathbf{v}(t)$ *is defined.*

This definition implies that $I = c =$ constant along the given solution. This defines a $6n$-dimensional *integral manifold,*

$$I^{-1}(0) = \{(\mathbf{x}, \mathbf{v}, t) \in \mathbf{R^{6n+1}} | I = c\},$$

on which the solutions will lie.

Thus, an integral constrains the motion of the mass particles and can be used to reduce the dimension of the space of $6n+1$ coordinates, $x_{k\ell}, v_{k\ell}, t, k = 1, 2, \ldots, n; \ell = 1, 2, 3$ by 1, by solving for one of the coordinates as a function of the $6n$ remaining coordinates, at least implicitly. For notation we refer to the $6n$-dimensional real space of coordinates $(\mathbf{x}, \mathbf{v}) \in \mathbf{R^{3n}} \times \mathbf{R^{3n}}$, as the *phase space*, and $(\mathbf{x}, \mathbf{v}, t) \in \mathbf{R^{3n}} \times \mathbf{R^{3n}} \times \mathbf{R^1}$ as the *extended phase space.*

When two or more integrals $I_1(\mathbf{x}, \mathbf{v}, t), I_2(\mathbf{x}, \mathbf{v}, t)$ exist for (1.1), they are called *independent* if the gradient vectors $\partial_{\mathbf{x},\mathbf{v},t} \equiv (\partial_{\mathbf{x}_1}, \ldots, \partial_{\mathbf{x}_n}, \partial_{\mathbf{v}_1}, \ldots, \partial_{\mathbf{v}_n}, \partial_t)$ of I_1 and I_2 are independent. This implies that the rank of the $2 \times (6n+1)$ matrix

$$\frac{\partial(I_1, I_2)}{\partial(\mathbf{x}, \mathbf{v}, t)}$$

is in general 2.

Equation (1.1) has a set of 10 independent algebraic integrals. These are given by the three classical conservation laws of linear momentum, energy, and angular momentum. We will derive these now.

First, we derive the conservation of linear momentum. To do this, we add up the right side of (1.1),

$$S = \sum_{k=1}^{n}\sum_{j=1}^{n} \frac{Gm_j m_k}{r_{jk}^2}\frac{\mathbf{x}_j - \mathbf{x}_k}{r_{jk}},$$

$j \neq k$. $S = 0$ is verified, since each term $\mathbf{x}_j - \mathbf{x}_k$ occurs with its negative, and mutual cancellations occur for all the terms. This implies

$$\sum_{k=1}^{n} m_k \ddot{\mathbf{x}}_k = 0.$$

Setting

$$M = \sum_{k=1}^{n} m_k, \qquad \boldsymbol{\rho} = M^{-1}\sum_{k=1}^{n} m_k \mathbf{x}_k,$$

where $\boldsymbol{\rho} = (\rho_1, \rho_2, \rho_3) \in \mathbf{R}^3$ is the *center of mass* vector of the particles, then $\ddot{\boldsymbol{\rho}} = \mathbf{0}$. This yields

$$\boldsymbol{\rho} = \mathbf{c}_1 t + \mathbf{c}_2, \tag{1.4}$$

$|t| < \delta$, where $\mathbf{c}_1, \mathbf{c}_2$ yield six constants which are uniquely determined from the initial conditions $\mathbf{x}_k(t_0), \mathbf{v}_k(t_0)$. Equation (1.4) expresses the law of the conservation of linear momentum: *The center of mass moves uniformly in a straight line.*

The origin of the Cartesian coordinate system x_1, x_2, x_3 for the motion of P_k can be shifted to the center of mass by setting $\bar{x}_j = x_j - \rho_j$. This does not alter the form of (1.1) since $\ddot{\rho}_j = 0$, and we can replace $\mathbf{x}_j$ by $\bar{\mathbf{x}}_j$. Thus, without loss of generality, we can assume $\boldsymbol{\rho} = \mathbf{0}$, which implies

$$\sum_{k=1}^{n} m_k \mathbf{x}_k = \mathbf{0}, \tag{1.5}$$

and also by differentiation,

$$\sum_{k=1}^{n} m_k \mathbf{v}_k = \mathbf{0}. \tag{1.6}$$

It is verified that (1.5), (1.6) represent six independent algebraic integrals $I_k, k = 1, 2, \ldots, 6$.

Another independent algebraic integral is given by the conservation of energy H,

$$H = T - U, \tag{1.7}$$

where H is the total energy of the system of n particles, and

$$T = \frac{1}{2}\sum_{k=1}^{n} m_k |\mathbf{v}_k|^2 \tag{1.8}$$

is the *kinetic energy*. Thus, H is the sum of the potential and kinetic energies. It is an integral since one verifies by direct computation that $\frac{d}{dt}(T-U)=0$ using (1.2). The law of conservation of energy states that the *energy is constant along solutions.*

The remaining three integrals are given by the conservation of angular momentum. This is derived by forming the vector cross product $\mathbf{x}_k \times \ddot{\mathbf{x}}_k$ using (1.1) and summing over k, where it is verified that

$$\sum_{k=1}^{n} m_k(\mathbf{x}_k \times \ddot{\mathbf{x}}_k) = \sum_{k=1}^{n}\sum_{j=1}^{n} \frac{Gm_j m_k}{r_{jk}^3} \mathbf{x}_k \times \mathbf{x}_j = \mathbf{0}, \tag{1.9}$$

where $j \neq k$ and where we used the fact $\mathbf{x}_k \times \mathbf{x}_k = \mathbf{0}$. The double sum is zero since $\mathbf{x}_j \times \mathbf{x}_k = -\mathbf{x}_k \times \mathbf{x}_j$. Integrating the left-hand side of (1.9) yields

$$\sum_{k=1}^{n} m_k(\mathbf{x}_k \times \mathbf{v}_k) = \mathbf{c}, \tag{1.10}$$

where $\mathbf{c} = (c_1, c_2, c_3) \in \mathbf{R}^3$ is the vector constant of angular momentum. Equation (1.10) expresses the law of conservation of angular momentum.

The angular momentum can be viewed as a measure of the rotational motion of (1.1). This measure of the rotation is illustrated in an important theorem of Sundman.

Theorem 1.2 (Sundman) *If at time $t = t_1$ all the particles P_k collide at one point, then $\mathbf{c} = \mathbf{0}$.*

This is called *total collapse.* The fact $\mathbf{c} = \mathbf{0}$ means that the particles are able to all collapse to a single location. In a sense, this is enabled because with $\mathbf{c} = \mathbf{0}$, the rotation has been taken away from the motion of the particles. For the two-body problem for $n = 2$, collision between P_1, P_2, where $r_{12} = 0$, can occur only if $\mathbf{c} = \mathbf{0}$. Theorem 1.2 is not proven here. (See [204].)

We conclude our introduction to the n-body problem with a brief summary of the extension of a solution $\mathbf{x}(t), \mathbf{v}(t)$ of (1.2) which has initial values $\mathbf{x}(t_0) = \mathbf{x}_0, \mathbf{v}(t_0) = \mathbf{v}_0$ at $t = t_0$. We extend this general solution for $t > t_0$. Now, either the $6n$ coordinates remain smooth for all time $t > t_0$, or else there is a first time $t = t_1$ where there is a *singularity* for at least one of the coordinates, where all coordinates are smooth for $t_0 \leq t < t_1$. The extent to which the solution can be continued in t beyond t_1 depends on whether or not, during the course of the motion of the P_k, the right hand-side of (1.2) remains smooth. Let $r_{\min}(t) = \min\{r_{jk}(t)\}, j < k$. $r_{\min}(t)$ is the minimum of the $n(n-1)/2$ distances r_{jk}. It can be proved that if t_1 is finite, then $r_{\min}(t) \to 0$ as $t \to t_1$. This implies $U \to \infty$ as $t \to t_1$. (See [204].)

In this case we say that there is a singularity of the solution at $t = t_1$. Surprisingly, this does not necessarily imply that a collision between the particles has to take place. This is called a *noncollision singularity*. The particles can get very close to each other and move in a complicated way so that the potential increases without bound. This question is a subtle one and is not considered in this book, as it is not the focus. However, we briefly summarize some key results on the nature of the singularity if $r_{\min} \to 0$ as $t \to t_1$.

If $n = 2$, then as $t \to t_1$ a collision must occur between m_1 and m_2. From Theorem 1.2, the condition that $\mathbf{c} = \mathbf{0}$ implies that the particles m_1, m_2 lie along a line, and as $r_{12} \to 0$ the collision can be *regularized*. This means that the solution can be smoothly continued to $t \geq t_1$ by a change of coordinates and time t. This is carried out in detail in sections 1.4 and 1.6. This means physically that m_1, m_2 perform a smooth bounce at collision and their motion for $t \geq t_1$ falls back along the same line on which they collided. Since $\mathbf{c} = \mathbf{0}$, three dimensions can be eliminated from the six-dimensional phase space. This means that the set of all collision orbits has a lower dimension than that of the phase space. The fact they have a lower dimension means that the total volume they make up is actually a set of relative zero volume in the full phase space. For example, in the two-dimensional Euclidean plane, the volume is *area* and all one-dimensional curves have zero area. The generalized volume of the phase space we use is called *measure*, which will mean *Lebesgue measure* [52]. Thus, the set of collision orbits in the two-body problem is a set of measure zero in the phase space. There is a natural way to assign a measure μ to the phase space $(\mathbf{x_1}, \mathbf{x_2}, \dot{\mathbf{x}}_1, \dot{\mathbf{x}}_2) \in \mathbf{R}^{12}$ of the three-dimensional two-body problem by setting

$$d\mu = dx_{11}dx_{12}dx_{13}\dots dx_{33}d\dot{x}_{11}d\dot{x}_{12}d\dot{x}_{13}\dots d\dot{x}_{33}.$$

This defines a twelve-dimensional volume element and generalizes in the natural way to the n-body problem, $n > 2$.

When $n = 3$ the situation is much more complicated. Two cases are considered. The first case is when $\mathbf{c} \neq \mathbf{0}$. By Theorem 1.2, simultaneous collision between all three particles cannot occur, and *only* binary collisions can occur between $P_k, P_\ell, k < \ell, k, \ell = 1, 2, 3$. As is shown in section 1.4, the set of all possible orbits leading to binary collisions in the eighteen-dimensional phase space is 16. Thus, they comprise a set of lower dimension and are of measure zero in the phase space. All these collisions in three dimensions can also be regularized by a transformation of position, velocity, and time, as described in Section 1.4, due to Sundman. This regularization uses a uniform time variable λ for all the collisions. After this transformation is applied, any solution can be continued through binary collision. Since only binary collisions can occur, Sundman was able to prove that any solution of the three-body problem can be extended for all time and also be explicitly represented as a series expansion [216].

Theorem 1.3 (Sundman) *Any solution of the general three-body problem with* $\mathbf{c} \neq \mathbf{0}$ *can be continued for all time and represented as a series expansion in the time variable* λ *that represents the entire motion.*

This theorem is of important historical significance since finding a way to explicitly express the solutions of the n-body problem was an outstanding problem for many years prior to that time. Theorem 1.3 does not actually solve the three-body problem since it does not describe the actual dynamics. Nevertheless, it does represent a milestone.

To underscore the importance of trying to solve the n-body problem, King Oscar II of Sweden and Norway had established a prize for this in the latter 19th century. The prize was for finding a series expansion for the coordinates of the n-body problem valid for all time. Although Sundman indeed solved this for $n = 3$ in 1913, he did not receive the prize. Instead it went to Poincaré much earlier–in 1889. Even though Poincaré did not solve the problem, he was given the prize due to the large impact his work had on the entire field of dynamics. For a detailed proof of Theorem 1.3, see [204].

When $\mathbf{c} = \mathbf{0}$, the total collapse of the three particles can occur. The ability to regularize collision is related to the uniformity of collision among the three particles. As we saw in the case of two-body collision, collision is uniform when the two particles perform a smooth bounce. The fact that they collide at $t = t_1$ means that $U \to \infty$ as $t \to t_1$, and conversely. A noncollision singularity between three particles would imply that near collision between the three particles a smooth regularized flow would not be possible to achieve in general. On the other hand, nonregularizability of collision does not imply the existence of a noncollision singularity. Nevertheless, nonregularizability of collision is a necessary condition for noncollision singularity states, and its existence plays an important role.

The question of whether or not triple collision was regularizable was solved by McGehee [151] for the collinear three-body problem. In this case all three mass points lie on a line. The phase space is six-dimensional since three position coordinates are needed, one for each mass point, and correspondingly three velocity coordinates. The method of McGehee's proof is to introduce a change of coordinates and time so that the triple collision state is transformed into a lower dimensional manifold. This surface then corresponds to the state when all three masses simultaneously collide and is called the *McGehee triple collision manifold.* The nonregularizability of triple collision is realized by the property that solutions approaching near triple collision, and hence near to the triple collision manifold, are led to widely divergent paths for small changes in their orbits. In the collinear three-body problem it can be seen that as $U \to \infty, r_{\text{min}} \to 0$ implies that the three particles do in fact collide. The proof of this for the general three-body problem was given by Painlevé in the late 19th century. Thus, in this case nonregularizability of collision does not imply the existence of a noncollision singularity.

Thus, if a noncollision singularity occurs, it would have to occur in at least the four-body problem. This is the Painlevé conjecture. More precisely, the conjecture states that *for* $n > 3$, *there exist solutions with noncollision singularities.* This was stated in 1895. [72].

Inspired by Painlevé, von Zeipel proved the following interesting theorem in 1908 [222].

Theorem 1.4 (von Zeipel) *If a noncollision singularity occurs in the n-body problem,* $n > 3$, *then there would exist a solution which would become unbounded in finite time.*

This would seem to violate the fact that the speed of light is the maximal velocity for particles of mass, however, in our case P_k are of zero dimension, and thus the distance between P_k can become arbitrarily small and the velocities arbitrarily high.

The solution of the Painlevé conjecture was solved by Xia for the spatial five-body problem and published in 1992 [227].

Theorem 1.5 (Xia) *There exist noncollision singularities in the spatial five-body problem.*

For an historical exposition of the Painlevé conjecture and related topics, see [72]. The existence of a noncollision singularity in the spatial four-body problem is an open problem.

The focus of this book will be on solutions which generally do not collide. In fact, solutions will be studied for the three-body problem which do not collide and where the motion is chaotic. We will study this near special families of periodic orbits in chapter 2, and with the process of capture defined in chapter 3.

1.2 PLANAR THREE-BODY PROBLEM

In later sections we will consider the planar circular restricted three-body problem as one of our main models. We derive it by first considering the general planar three-body problem, using Jacobi coordinates [191, 219].

The planar three-body problem is obtained from (1.2) by setting $n = 3$ and assuming $\mathbf{x}_k \in \mathbf{R}^2, \mathbf{v}_k \in \mathbf{R}^2, k = 1, 2, 3$.

The center of mass $\boldsymbol{\rho} = \mathbf{0}$, in accordance with (1.5), where $\boldsymbol{\rho} = M^{-1}\sum_{k=1}^{3} m_k\mathbf{x}_k, M = m_1 + m_2 + m_3, \sum_{k=1}^{3} m_k\mathbf{x}_k = \mathbf{0}$. It is useful to write out (1.1),

$$\begin{aligned} m_1\ddot{\mathbf{x}}_1 &= \frac{Gm_1m_2}{r_{12}^3}(\mathbf{x}_2 - \mathbf{x}_1) + \frac{Gm_3m_1}{r_{13}^3}(\mathbf{x}_3 - \mathbf{x}_1), \\ m_2\ddot{\mathbf{x}}_2 &= \frac{Gm_1m_2}{r_{12}^3}(\mathbf{x}_1 - \mathbf{x}_2) + \frac{Gm_3m_2}{r_{23}^3}(\mathbf{x}_3 - \mathbf{x}_2), \\ m_3\ddot{\mathbf{x}}_3 &= \frac{Gm_1m_3}{r_{13}^3}(\mathbf{x}_1 - \mathbf{x}_3) + \frac{Gm_3m_2}{r_{23}^3}(\mathbf{x}_2 - \mathbf{x}_3). \end{aligned} \tag{1.11}$$

This is a system of six second order differential equations. With the constraint $\boldsymbol{\rho} = \mathbf{0}$ we can eliminate one of the vector variables $\mathbf{x}_i$, resulting in four second order differential equations. Equations (1.11) are transformed to *Jacobi coordinates.* Set

$$\mathbf{q} = \mathbf{x}_2 - \mathbf{x}_1, \qquad \mathbf{Q} = \mathbf{x}_3 - \boldsymbol{\beta},$$

where $\boldsymbol{\beta} = \nu^{-1}(m_1\mathbf{x}_1 + m_2\mathbf{x}_2), \nu = m_1 + m_2$, is the center of mass vector of the binary pair m_1, m_2. $\mathbf{q}$ is the relative vector of P_2 with respect to P_1, and $\mathbf{Q}$ is the vector from $\boldsymbol{\beta}$ to P_3. (See Figure 1.1.) $\mathbf{q}, \mathbf{Q}$ are called Jacobi coordinates.

The transformation of (1.11) to Jacobi coordinates is now carried out. Each term $\mathbf{x}_k - \mathbf{x}_j, k > j, j, k = 1, 2, 3$, is transformed. First, by definition $\mathbf{q} = \mathbf{x}_2 - \mathbf{x}_1$. Next, $\mathbf{x}_3 - \mathbf{x}_1 = \mathbf{Q} + m_2\nu^{-1}\mathbf{q}$. This follows since $\mathbf{x}_3 - \mathbf{x}_1 = \mathbf{Q} + \boldsymbol{\beta} - \mathbf{x}_1 = \mathbf{Q} + \nu^{-1}(m_1\mathbf{x}_1 + m_2\mathbf{x}_2) - \mathbf{x}_1 = \mathbf{Q} + m_2(\mathbf{x}_2 - \mathbf{x}_1)\nu^{-1} = \mathbf{Q} + m_2\nu^{-1}\mathbf{q}$. In a similar way, $\mathbf{x}_3 - \mathbf{x}_2 = \mathbf{Q} - m_1\nu^{-1}\mathbf{q}$. Substituting the expressions for $\mathbf{x}_k - \mathbf{x}_j$ into (1.11), dividing the first differential equation by m_1, the second by m_2, and subtracting these two differential equations yields

$$\ddot{\mathbf{q}} = -\frac{G\nu}{|\mathbf{q}|^3}\mathbf{q} + Gm_3\left[\frac{\mathbf{Q} - m_1\nu^{-1}\mathbf{q}}{r_{23}^3} - \frac{\mathbf{Q} + m_2\nu^{-1}\mathbf{q}}{r_{13}^3}\right], \tag{1.12}$$

where $r_{12} = |\mathbf{q}|, r_{13} = |\mathbf{Q} + m_2\nu^{-1}\mathbf{q}|, r_{23} = |\mathbf{Q} - m_1\nu^{-1}\mathbf{q}|$. To obtain the differential equation for $\mathbf{Q}$, we use the third equation of (1.11). We need for this another relationship for $\mathbf{Q} = \mathbf{x}_3 - \boldsymbol{\beta} = \mathbf{x}_3 - \nu^{-1}(m_1\mathbf{x}_1 + m_2\mathbf{x}_2) = \mathbf{x}_3 - \nu^{-1}(-m_3x_3) = (1 + \nu^{-1}m_3)\mathbf{x}_3 = \nu^{-1}M\mathbf{x}_3$. This implies $\mathbf{x}_3 = \nu M^{-1}\mathbf{Q}$. Multiplying the third equation of (1.11) by $m_3^{-1}\nu^{-1}M$ yields

$$\ddot{\mathbf{Q}} = -\frac{GMm_1\nu^{-1}}{r_{13}^3}(\mathbf{Q} + m_2\nu^{-1}\mathbf{q}) - \frac{GMm_2\nu^{-1}}{r_{23}^3}(\mathbf{Q} - m_1\nu^{-1}\mathbf{q}). \tag{1.13}$$

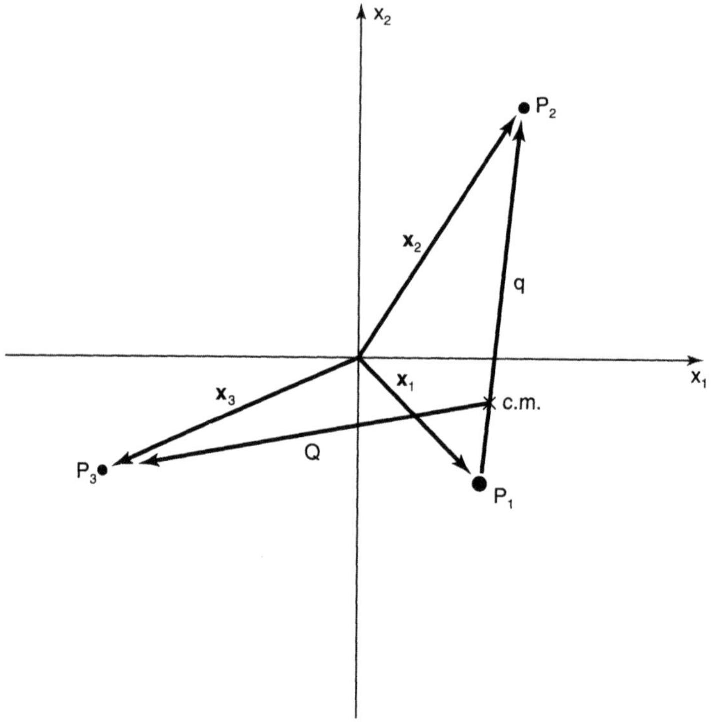

Figure 1.1 Jacobi coordinates.

Equations (1.12), (1.13) represent (1.11) in Jacobi coordinates, which are four second order differential equations for $(\mathbf{q}, \mathbf{Q}) = (q_1, q_2, Q_1, Q_2) \in \mathbf{R^4}$.

The kinetic energy T of this system given by (1.8) for $n = 3$ takes a nice form if we set

$$\dot{\mathbf{q}} = k_2^{-1}\mathbf{p}, \qquad \dot{\mathbf{Q}} = k_1^{-1}\mathbf{P}, \tag{1.14}$$

where

$$k_1 = m_3 \nu M^{-1}, \qquad k_2 = m_1 m_2 \nu^{-1}.$$

It is noted that if $m_3 = 0$, then $\mathbf{P} = \mathbf{0}$.

Lemma 1.6 *The kinetic energy of the system (1.12), (1.13) is given by*

$$T = \frac{1}{2}(k_2^{-1}|\mathbf{p}|^2 + k_1^{-1}|\mathbf{P}|^2) \tag{1.15}$$

and the total energy $\mathcal{H}$ of the system is

$$\mathcal{H} = T - U,$$
$$U = \frac{Gm_1m_2}{|\mathbf{q}|} + \frac{Gm_2m_3}{r_{23}} + \frac{Gm_1m_3}{r_{13}}. \tag{1.16}$$

Proof. The form of $\mathcal{H}$ follows from (1.15) together with (1.7). To transform (1.8), we use (1.14). It is verified that from $\mathbf{x}_3 - \mathbf{x}_2 = \mathbf{Q} - m_1\nu^{-1}\mathbf{q}$, $\mathbf{x}_3 = M^{-1}\nu\mathbf{Q}$,

$$\begin{aligned} \mathbf{x}_1 &= -m_3M^{-1}\mathbf{Q} - m_2\nu^{-1}\mathbf{q}, \\ \mathbf{x}_2 &= -m_3M^{-1}\mathbf{Q} + m_1\nu^{-1}\mathbf{q}. \end{aligned}$$

Then, (1.14) implies

$$\begin{aligned} \dot{\mathbf{x}}_1 &= -\nu^{-1}\mathbf{P} - m_1^{-1}\mathbf{p}, \\ \dot{\mathbf{x}}_2 &= -\nu^{-1}\mathbf{P} + m_2^{-1}\mathbf{p}, \\ \dot{\mathbf{x}}_3 &= m_3^{-1}\mathbf{P}. \end{aligned}$$

Substituting these relationships into (1.8) yields (1.15) after simplification. □

It is immediately verified that (1.12), (1.13) can be written in Hamiltonian form,

$$\begin{aligned} \dot{\mathbf{q}} &= \mathcal{H}_{\mathbf{p}}, \quad \dot{\mathbf{p}} = -\mathcal{H}_{\mathbf{q}}, \\ \dot{\mathbf{Q}} &= \mathcal{H}_{\mathbf{P}}, \quad \mathbf{P} = -\mathcal{H}_{\mathbf{Q}}, \end{aligned} \tag{1.17}$$

where $\mathcal{H}_{\mathbf{p}} \equiv \partial\mathcal{H}/\partial\mathbf{p}$.

Jacobi coordinates are particularly well suited to studying versions of the three-body problem where P_1, P_2 are performing a given binary motion and the mass of P_3 is infinitesimally small. This situation would occur, for example, if P_3 were considered to be a small object such as a spacecraft, comet, or asteroid, if P_1 were the Sun, and if P_2 were Jupiter. For all practical purposes, m_3 has negligible mass. Setting $m_3 = 0$ reduces (1.12) to

$$\ddot{\mathbf{q}} = \frac{G\nu}{|\mathbf{q}|^3}\mathbf{q}, \tag{1.18}$$

which represents the standard differential equation for two-body problem for the motion of P_1, P_2 in relative coordinates centered at P_1. Equation (1.18) can be explicitly solved. Equation (1.13) then describes the motion of P_3 in the gravitational field generated by the motion of the particles P_1, P_2 defined by (1.18).

Before proceeding to the restricted three-body problem, the solutions of (1.18) are derived.

1.3 TWO-BODY PROBLEM

The *Keplerian two-body problem* is defined by (1.18) and is in the coordinates shown in Figure 1.2.

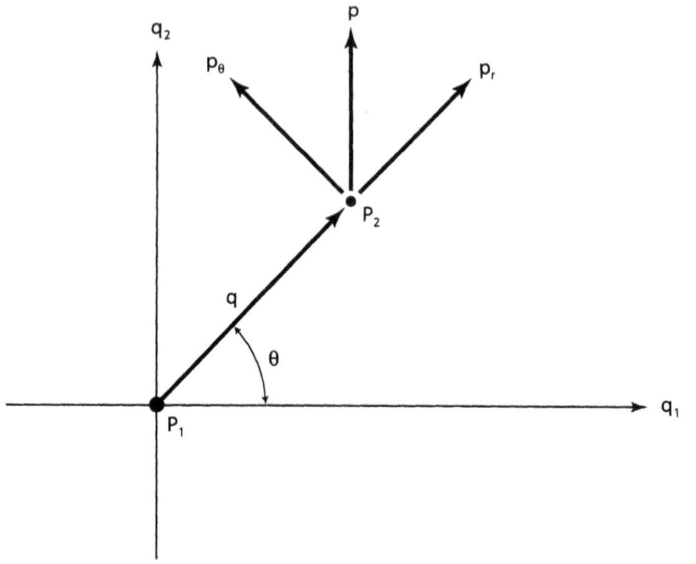

Figure 1.2 Polar coordinates and momenta.

The total energy of the system is given by the real-valued function, $H(p,q) : \mathbf{R^4} \to \mathbf{R^1}$,

$$H = \frac{1}{2}|\mathbf{p}|^2 - \frac{G\nu}{|\mathbf{q}|}, \tag{1.19}$$

where $\mathbf{p} = \dot{\mathbf{q}} \in \mathbf{R^2}$. It is noted that the coefficient of $\frac{1}{2}|\mathbf{p}|^2$ is 1 and does not contain m_1, m_2 as in the general form of the energy (1.7). This is because (1.7) is written in terms of inertial coordinates, and (1.18) is in relative coordinates. $\mathbf{p}$ are referred to as the *momentum* variables, which can be viewed as linear momentum of a unit mass since $\mathbf{p} = 1\dot{\mathbf{q}}$. Equation (1.18) is the first order system

$$\dot{\mathbf{q}} = \mathbf{p}, \quad \dot{\mathbf{p}} = -\frac{G\nu}{|\mathbf{q}|^3}\mathbf{q}, \tag{1.20}$$

or, in Hamiltonian form, using (1.19),

$$\dot{\mathbf{q}} = H_{\mathbf{p}}, \quad \dot{\mathbf{p}} = -H_{\mathbf{q}}. \tag{1.21}$$

H is called the *Hamiltonian function.* The solutions of (1.20) are well known and depend on the value $H = h$ of the energy. If $h < 0$, the curve $\mathbf{q}(t)$ is an ellipse with a focus at P_1. For $h = 0, \mathbf{q}(t)$ is a parabola, and for $h > 0, \mathbf{q}(t)$ is a hyperbola, where both the parabola and hyperbola have foci at the origin, or alternatively P_1. These solutions are derived in this section and section 1.6.

The angular momentum is given by

$$\mathbf{c} = \mathbf{q} \times \mathbf{p} = q_1 p_2 - q_2 p_1. \tag{1.22}$$

We prove that

$$|\mathbf{c}| = c = r^2\dot{\theta}, \tag{1.23}$$

$r = |\mathbf{q}|, \theta$ is the polar angle shown in Figure 1.2. Referring to Figure 1.2, $\mathbf{p}$ can be decomposed into its tangential and radial components, $\mathbf{p}_\theta, \mathbf{p_r}$, respectively, $\mathbf{p} = \mathbf{p}_r + \mathbf{p}_\theta = p_r\mathbf{e}_r + p_\theta\mathbf{e}_\theta$, where $\mathbf{e}_r, \mathbf{e}_\theta$ are unit vectors in the r, θ directions, respectively.

Differentiating $\mathbf{q} = r\mathbf{e}_r$ implies $\dot{\mathbf{q}} = \mathbf{p} = r\dot{\mathbf{e}}_r + \dot{r}\mathbf{e}_r$, where $\dot{\mathbf{e}}_r = \dot{\theta}\mathbf{e}_\theta$, which follows since $\mathbf{e}_r$ rotates with constant circular velocity $\dot{\theta}$. (Similarly, $\dot{\mathbf{e}}_\theta = -\dot{\theta}\mathbf{e}_r$.) Thus, $\mathbf{c} = \mathbf{q}\times\mathbf{p} = r^2\dot{\theta}\mathbf{e}_r\times\mathbf{e}_\theta$, verifying (1.23). The constancy of $\mathbf{c}$ gives the law discovered by Kepler that as m_2 moves in an elliptic orbit about m_1, it traces out equal areas in equal times. This is *Kepler's first law.* This follows since the change of area $A(t)$ that is swept out by m_2 in time t is approximated by $\Delta A = \frac{1}{2}r^2\dot{\theta}\Delta t$ for $\Delta t \ll 1$, since the base of the triangle in Figure 1.3 has length $rd\theta \approx r\dot{\theta}\Delta t$.

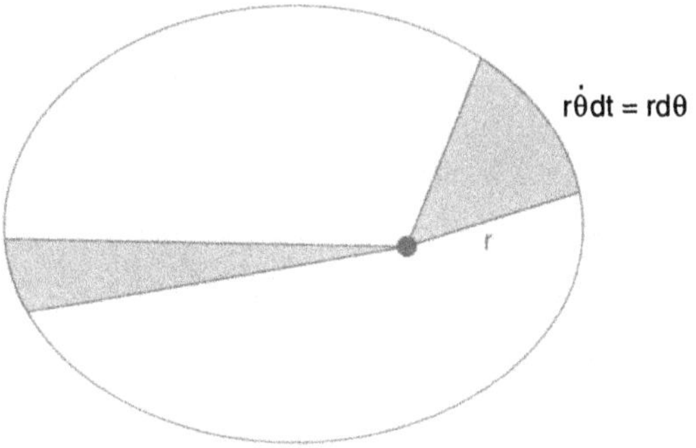

Figure 1.3 Sectorial area.

Therefore, as $\Delta t \to 0, \dot{A} = \frac{1}{2}c =$ constant. $\dot{A}$ is called the *sectorial area.*

The constancy of $\mathbf{c}$ also implies that all orbits are planar. Letting $\mathbf{q} \in \mathbf{R^3}, \mathbf{p} \in \mathbf{R^3}$, then $\mathbf{c}\times\mathbf{q} = (\mathbf{q}\times\mathbf{p})\times\mathbf{q} = \mathbf{0}$, which means that $\mathbf{q}(t)$ is perpendicular to $\mathbf{c}, t \in \mathbf{R^1}$. Thus, in (1.18) it is sufficient to assume $\mathbf{q} \in \mathbf{R^2}$.

Using a special regularizing transformation defined in the following sections, the solutions of (1.21) are explicitly determined and are nonsingular in the entire phase space. However, we derive here for future reference the solution for (1.18) or (1.21) with $h < 0$. Equation (1.18) is solved by transforming it to polar coordinates r, θ. Differentiating the previously stated polar representation for $\dot{\mathbf{q}}$ yields

$$\ddot{\mathbf{q}} = (\ddot{r} - r\dot{\theta}^2)\mathbf{e}_r + (2\dot{r}\dot{\theta} + r\ddot{\theta})\mathbf{e}_\theta.$$

On the other hand, (1.18) implies $\ddot{\mathbf{q}} = U_\mathbf{q}, U = G\nu/r$ and since $U_\mathbf{q}$ represents a central force field which is radially directed,

$$U_\mathbf{q} = U_r\mathbf{e}_r.$$

Thus, equating coefficients in $\ddot{\mathbf{q}}$ yields

$$\ddot{r} - r\dot{\theta}^2 = U_r, \qquad 2\dot{r}\dot{\theta} + r\ddot{\theta} = 0.$$

Since $r^2\dot{\theta} = c =$ constant, then knowing $r(t)$ we can determine $\dot{\theta}(t)$, and hence $\theta(t)$ by quadratures. Therefore, it is sufficient to solve the first differential equation for $r(t)$, which we can write as

$$\ddot{r} = V_r, \tag{1.24}$$

where

$$V = U - \frac{c^2}{2r^2}$$

is called the *effective potential energy.* Thus, we have proved

Lemma 1.7 *Equation (1.18) can be reduced to solving (1.24).*

Lemma 1.8 *The total energy associated with (1.24) is given by*

$$H = \frac{1}{2}\dot{r}^2 - V. \tag{1.25}$$

Proof. In polar coordinates, (1.19) becomes

$$H = \frac{1}{2}(\dot{r}^2 + r^2\dot{\theta}^2) - U = \frac{1}{2}(\dot{r}^2 + r^2\dot{\theta}^2) - V - \frac{c^2}{2r^2},$$

and $c = r^2\dot{\theta}$ yields (1.25). □

Since H is a real constant, then (1.25) can be used to solve (1.24) by solving for $\dot{r}$, and using quadrature,

$$\int dt = \int \frac{dr}{\sqrt{2(H - V(r))}},$$

which implicitly yields $r = r(t)$.

This integral equation is used to solve explicitly for $r = r(\theta)$. Using this expression and the one for c, we obtain

$$\frac{d\theta}{dt} = \frac{d\theta}{dr}\frac{dr}{dt} = \frac{d\theta}{dr}\sqrt{2(H - V(r))}.$$

Thus

$$\theta = \int \frac{(c/r^2)dr}{\sqrt{2(H - V(r))}}. \tag{1.26}$$

θ is called the *true anomaly.*

Lemma 1.9 *Equation (1.26) implies*

$$r = \frac{p}{1 + e\cos\theta}, \tag{1.27}$$

where

$$p = c^2/k, \quad e = \left(1 + \frac{2Hc^2}{k^2}\right)^{\frac{1}{2}}, \quad k = G\nu.$$

Proof. Integration of (1.26) gives

$$\theta(r) = \cos^{-1}\left[\frac{cr^{-1} - c^{-1}k}{\left(2H + \frac{k^2}{c^2}\right)^{\frac{1}{2}}}\right] + \theta(r_0)$$

which gives (1.27), where $\theta(r_0) = 0$ and $r_0 = r_p, r_p$ is the periapsis radius, which is the radius of closest approach. □

Equation (1.27) represents an ellipse, where $H = h < 0, e < 1$. e is the *eccentricity*, and p is the *semilatus rectum.* (See Figure 1.4.) The focus of the ellipse is at the origin, where m_1 is located. This is *Kepler's second law*, that the attracting mass occupies a focus of the particle's orbit. r_a is the furthest point on the ellipse to the origin, called the *apoapsis.*

Different relationships between the parameters a, e, p, etc. can be derived, where a is the *semimajor axis.* For example, $g = ae$ is the distance from the center of the ellipse to the focus. This implies $f^2 = a^2 - g^2 = a^2(1 - e^2)$. Also, $a = p/(1 - e^2)$. It is verified that $r_p = a(1 - e), r_a = a(1 + e)$. For a

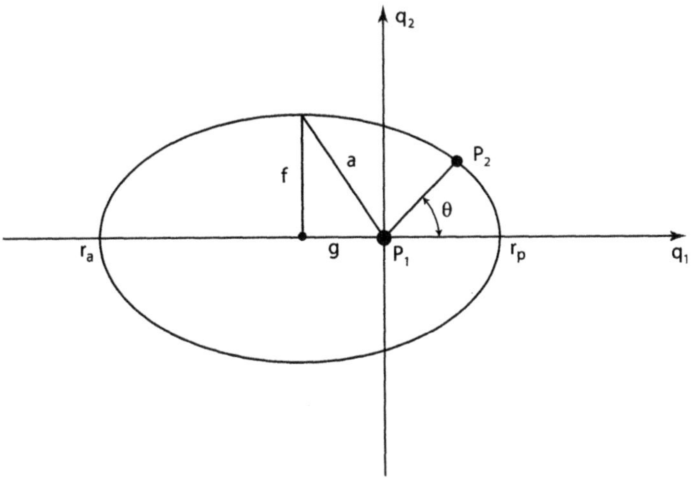

Figure 1.4 Geometry of the Kepler ellipse.

circular orbit, it is checked that $e = 0$, which implies $r = p = a =$ constant. Equation (1.27) is often expressed as

$$r = \frac{a(1-e^2)}{1+e\cos\theta}. \tag{1.28}$$

An important relationship is the connection between H and a. From $e = (1+2Hc^2/k^2)^{\frac{1}{2}}, a = p/(1-e^2)$, we obtain

$$H = -\frac{k}{2a}. \tag{1.29}$$

Thus, a is a measure of the amount of energy contained in the Kepler motion.

Kepler's third law says that the square of the period of motion, P, along the ellipse (1.27) is proportional to the cube of a. This is derived using $A(t)$. Since $\frac{dA}{dt} = \frac{1}{2}c$, then

$$A(P) = \frac{1}{2}cP = \text{area of the ellipse} = \pi a f.$$

Therefore, since $f = a\sqrt{1-e^2}$,

$$P = 2\pi c^{-1}a^2\sqrt{1-e^2},$$

which by substitution of (1.29) and $e = (1+2Hc^2/k^2)^{\frac{1}{2}}$, yields

$$P^2 = (2\pi)^2 k^{-1} a^3, \tag{1.30}$$

which is Kepler's third law.

Finally, the *mean anomaly* is given by an angle $\mathcal{M}$ varying during one revolution from 0 to 2π with constant angular velocity. Since $k^{\frac{1}{2}}P/a^{\frac{3}{2}} = 2\pi$, then

$$\mathcal{M} = \frac{k^{\frac{1}{2}}}{a^{\frac{3}{2}}}t.$$

The *eccentric anomaly* E is defined as

$$\mathcal{M} = E - e\sin E.$$

It is seen in the next section that E is desirable to use in place of t to describe the Kepler motion.

When $H \geq 0$, an analogous description of the parabolic and hyperbolic motion can be made (see [214]). In the next section all values of H are treated in a uniform manner.

1.4 REGULARIZATION OF COLLISION

Equation (1.20) is singular when $\mathbf{q} = \mathbf{0}$, where, as $\mathbf{q} \to \mathbf{0}, U = k|\mathbf{q}|^{-1} \to \infty$. A solution $\mathbf{q}(t), \mathbf{p}(t)$ of (1.20), where $\mathbf{q}(t) \to \mathbf{0}, |\mathbf{p}(t)| \to \infty$ as $t \to t_0$, is called a *collision orbit*, where collision is given by $\mathbf{q} = 0$ at $t = t_0$ between m_1 and m_2. Let $\varphi(t) = (\mathbf{q}(t), \mathbf{p}(t)) \in \mathbf{R^4}$ represent a collision orbit. We show how to continue a solution through collision in a smooth fashion.

Let

$$\mathbf{q} = \mathbf{q}(\mathbf{u}, \mathbf{w}), \qquad \mathbf{p} = \mathbf{p}(\mathbf{u}, \mathbf{w}), \qquad t = t(\mathbf{u}, \mathbf{w}, s) \tag{1.31}$$

be a transformation, where $\mathbf{u} \in \mathbf{R^m}, \mathbf{w} \in \mathbf{R^m}, s \in \mathbf{R^1}, m \geq 2$, which transforms (1.20) into a system

$$\mathbf{u}' = \mathbf{f}(\mathbf{u}, \mathbf{w}), \qquad \mathbf{w}' = \mathbf{g}(\mathbf{u}, \mathbf{w}), \tag{1.32}$$

$' \equiv \frac{d}{ds}$, and which transforms $\varphi(t)$ into $\boldsymbol{\zeta}(s) = (\mathbf{u}(s), \mathbf{w}(s))$. We assume that $s = s_0 < \infty$ corresponds to collision, where $\mathbf{q} = 0$ corresponds to $\mathbf{u} = \mathbf{u}_0$ and $\mathbf{w} \in \mathcal{C}, \mathcal{C}$ is a bounded subset of $\mathbf{R^m}$. $\mathbf{u}$ is the transformed position vector, and $\mathbf{w}$ is the transformed velocity vector. Assume that a given collision orbit $\varphi(t), t \neq t_0$, is mapped into $\boldsymbol{\zeta}(s), s \neq s_0$, and as $s \to s_0, \mathbf{w}(s) \to \mathbf{w}_0 \in \mathcal{C}$.

Definition 1.10 *If (1.32) is smooth in a neighborhood of the set* $A = \{(\mathbf{u}_0, \mathbf{w}), \mathbf{w} \in \mathcal{C}\}$, *which are assumed to be nonequilibrium points of (1.32), then (1.20) is* regularizable *at* $\mathbf{q} = 0$, *and (1.31) is called a* regularization.

Thus, by this definition any collision solution $\boldsymbol{\zeta}(s)$ can be extended smoothly as a function of s through $s = s_0$, for s in a neighborhood of s_0. Also, any collision solution $\boldsymbol{\zeta}(s, \tilde{\mathbf{u}}_0, \tilde{\mathbf{w}}_0)$ is a smooth function of $\tilde{\mathbf{u}}_0, \tilde{\mathbf{w}}_0$, where $\tilde{\mathbf{u}}_0, \tilde{\mathbf{w}}_0$ are in a neighborhood of A. In this case (1.31) is called a *local regularization.* If (1.32) is smooth for all $(\mathbf{u}, \mathbf{w}) \in \mathbf{R^4}$, then (1.31) is called a *global regularization.* This is not the most general definition of regularization, but it is sufficient for our presentation.

We consider a special regularization due to Levi-Civita [134]. In complex notation it is given by

$$q = z^2, \quad p = \frac{wz}{2|z|^2}, \tag{1.33}$$

where $q = q_1 + iq_2 \in \mathbb{C}, w = w_1 + iw_2 \in \mathbb{C}, i^2 = -1, |z|^2 = z\bar{z}, \bar{z} = z_1 - iz_2$. Thus $q = 0$ is mapped into $z = 0$.

(1.33) is written in *canonical form.* This means that it preserves the differential form $w = dp_1dq_1 + dp_2dq_2 = Re(\overline{dp}dq) = Re(\overline{dw}dz)$, where $Re(z) = z_1$.

Lemma 1.11 *ω is invariant under (1.33).*

Proof. We show that (1.33) satisfies $\bar{p}dq = \bar{w}dz$ under (1.33). Lemma 1.11 follows by applying the operator d to both sides of $\bar{p}dq = \bar{w}dz$ and noting that $d^2 = 0$. Assume $q = z^2$. Then

$$\bar{p}dq = 2\bar{p}zdz.$$

Thus, setting $\bar{w} = 2\bar{p}z$ yields the invariance of ω and implies

$$p = \frac{w}{2\bar{z}} = \frac{wz}{2|z|^2},$$

giving (1.33). □

The invariance of ω implies that the area elements of the phase space are preserved. By Liouville's theorem, this implies that the Hamiltonian form of the differential equations are preserved [14]. Thus, the form of (1.21) is preserved in the new coordinates. If

$$\Phi(\mathbf{z}, \mathbf{w}) = H(\mathbf{q}(\mathbf{z}), \mathbf{p}(\mathbf{z}, \mathbf{w})),$$

then

$$\dot{\mathbf{z}} = \Phi_{\mathbf{w}}, \ \mathbf{w} = -\Phi_{\mathbf{z}}, \tag{1.34}$$

$\mathbf{z} = (z_1, z_2) \in \mathbf{R^2}, \mathbf{w} = (w_1, w_2) \in \mathbf{R^2}$, and

$$\Phi(\mathbf{z}, \mathbf{w}) = \frac{1}{8}|\mathbf{z}|^{-2}|\mathbf{w}|^2 - k|\mathbf{z}|^{-2}. \tag{1.35}$$

We restrict H to the value $H = h$, so that (1.21) is defined on the three-dimensional energy level $H^{-1}(h) = \{(\mathbf{p}, \mathbf{q}) \in \mathbf{R^4} | H(\mathbf{p}, \mathbf{q}) = h\}$. Thus, (1.34) is defined on the set

$$\Phi^{-1}(h) = \{(\mathbf{z}, \mathbf{w}) \in \mathbf{R^4} | \Phi(\mathbf{z}, \mathbf{w}) = h\}.$$

The Levi-Civita transformation is augmented by the time transformation

$$dt = |\mathbf{q}|ds = |\mathbf{z}|^2 ds, \tag{1.36}$$

which as $|\mathbf{z}| \to 0$ stretches the new time variable s. This implies that (1.34) takes the form

$$\mathbf{z}' = |\mathbf{z}|^2 \Phi_{\mathbf{w}}, \qquad \mathbf{w}' = -|\mathbf{z}|^2 \Phi_{\mathbf{z}},$$

$' \equiv \frac{d}{ds}$, or equivalently

$$\mathbf{z}' = \tilde{\Phi}_{\mathbf{w}}, \qquad \mathbf{w}' = -\tilde{\Phi}_{\mathbf{z}}, \tag{1.37}$$

where $\tilde{\Phi} = (\Phi - h)|\mathbf{z}|^2$, or

$$\tilde{\Phi} = \frac{1}{8}|\mathbf{w}|^2 - h|\mathbf{z}|^2 - k \tag{1.38}$$

and $\mathbf{w}, \mathbf{z}$ are restricted to the set where $\tilde{\Phi} = 0$.

Theorem 1.12 *The Kepler problem given by (1.21) on the three-dimensional energy surface $H^{-1}(h), h \in \mathbf{R}^1$, is transformed by the Levi-Civita transformation (1.33) together with the time transformation (1.36) into the second order linear system*

$$\mathbf{z}'' - \frac{h}{2}\mathbf{z} = 0. \tag{1.39}$$

Collision for (1.21) is mapped onto the set $A = \{\mathbf{z}, \mathbf{w} | \mathbf{z} = 0, |\mathbf{w}|^2 = 8k\}, k = G\nu \neq 0$, where (1.39) is smooth and $\mathbf{w} = 4\mathbf{z}'$.

Proof. Equation (1.37) implies

$$\mathbf{z}' = \frac{1}{4}\mathbf{w}, \qquad \mathbf{w}' = 2h\mathbf{z},$$

which yields (1.39). Setting $\mathbf{z} = 0$ in (1.38) yields $|\mathbf{w}|^2 = 8k$, and hence the set A. Equation (1.39) is smooth at $\mathbf{z} = 0$. □

Lemma 1.13 *Equations (1.33), (1.36) represent a global regularization of (1.21).*

Proof. A does not represent a location where (1.39) has equilibrium points. Thus, any collision solution can be extended up to $\mathbf{z} = 0$ at a finite time $t = t_0 < \infty$, and then extended beyond collision for $t > t_0$. Thus, (1.33), (1.36) is a local regularization. It is a global regularization since (1.39) is smooth on the entire energy surface $\tilde{\Phi}^{-1}(0)$. □

In [23], examples of local regularizations are given. Ignoring canonical extensions and the time transformation, an example of one for (1.35) in the momentum coordinates is given by the so-called *Sundman transformation,*

$$\mathbf{p} \to \frac{\mathbf{p}}{|\mathbf{p}|^2} = \tilde{\mathbf{p}}.$$

$|\mathbf{p}| = \infty$ is mapped into $\tilde{\mathbf{p}} = \mathbf{0}$. Thus $|\mathbf{p}| = \infty$ is mapped into a finite point where a local regularization can be constructed by canonically extending this map to a map of the $\mathbf{q}$ variables and t. However, this is not a global regularization, since in the new coordinates $\tilde{\mathbf{p}}, \tilde{\mathbf{q}}, |\tilde{\mathbf{p}}| = \infty$ will be a singular point and it corresponds to the point $\mathbf{p} = \mathbf{0}$, which is smooth for (1.21). Thus, the Sundman transformation introduces another singularity into the phase space $(\tilde{\mathbf{p}}, \tilde{\mathbf{q}}) \in \mathbf{R}^4$.

It is remarked that A in Theorem 1.12 is equivalent to the set $\{\mathbf{z} = 0\} \times S^1$, where S^1 is a circle of radius $\sqrt{8k}$. Thus, collision for (1.21) has been reduced to the set product of a point and a circle.

It is noted that $c = 0$ is required for collision to occur. For the case of $h < 0$ corresponding to elliptic motion, for example, this implies that $e = 1$

by (1.27) for any solution leading to collision. By (1.29), for any finite $h < 0$, this implies that a is finite, and so is r_a, where $r_p = 0$. Thus as $e \uparrow 1$, or equivalently as $c \to 0$, a family of ellipses are obtained which get thinner and thinner and converge to a line representing a collision orbit, where $\theta =$ constant as follows from c. The existence of a smooth regularization means that m_2 collides with m_1 at a finite time, smoothly bounces off of m_1, moves away from m_1 along the same line, and then repeats the process. This is called a *consecutive collision orbit* [23, 219]. (See Figure 1.5.)

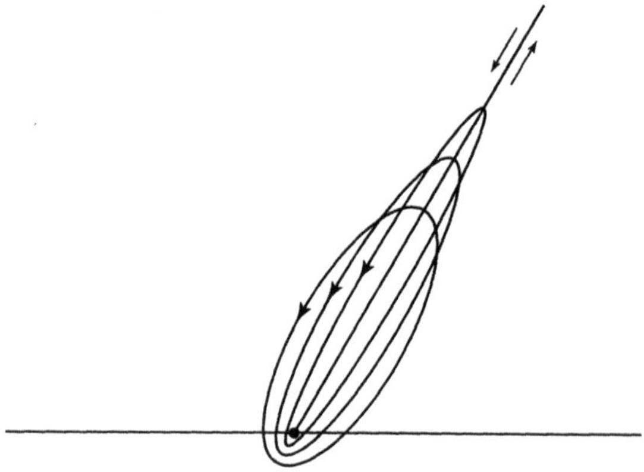

Figure 1.5 Evolution of ellipses to a consecutive collision orbit.

Lemma 1.14 *The set of collision orbits for (1.21) on each three-dimensional energy surface $H^{-1}(h)$ is topologically equivalent to the two-dimensional set $S^1 \times \mathbf{R^1}$.*

Proof. By Theorem 1.12, the collision manifold corresponds to the circle S^1, where $\mathbf{z} = \mathbf{0}$. Each point $\mathbf{w}_0$ on this circle can therefore be viewed as an initial value on the collision manifold of a collision orbit for $s = s_o$. Since the Levi-Civita regularization is global, each collision orbit is defined for all time $s \in \mathbf{R^1}$. Thus, the set of all collision orbits corresponds to the cylinder $C^2 = S^1 \times \mathbf{R^1}$. □

By Lemma 1.14, the set of all collision orbits in the four-dimensional phase space is then obtained by varying $h \in \mathbf{R^1}$, so we obtain the three-dimensional set $C^2 \times \mathbf{R^1}$. The dimension of this set agrees with the dimension of the set $\{\mathbf{c} = \mathbf{0}\}$.

Lemma 1.15 *The set of all collision orbits in the general planar three-body problem for $\mathbf{c} \neq \mathbf{0}$ is a set of dimension 11 in the 12-dimensional phase space.*

Proof. From the preceding discussion following the proof of Lemma 1.14, the set of all collision orbits in the planar two-body problem is a set of three dimensions in the four-dimensional phase space. Assume a binary collision occurs at time $t = t_0$ for (1.2), $\mathbf{x}_k \in \mathbf{R}^2, k = 1, 2, 3$. $t_0 < \infty$ is well defined since $\mathbf{c} \neq \mathbf{0}$ where only isolated binary collisions can occur. The collision manifold for any of the three possible binary collisions is then three dimensions. On the other hand, the center of mass of the binary collision can no longer be fixed, as conservation of linear momentum is not valid due to the presence of a third mass point. Thus, the binary collision depends on the location in phase space of the two masses relative to the third. This introduces eight free parameters due to the position and velocity of two mass points. This, together with the three dimensions of the collision manifold, yields eleven dimensions, in the twelve dimensional phase space. □

It is noted that in the proofs of Lemmas 1.14 and 1.15, the Levi-Civita regularization need not be used to deduce the dimension of the collision manifold and set of all collision orbits. Any regularization can be used locally or globally. In [216] a local regularization is in fact used in the case of the three-dimensional three-body problem. Consider the three-dimensional two-body problem defined by (1.20) with $\mathbf{p} \in \mathbf{R}^3, \mathbf{q} \in \mathbf{R}^3$. In that case it is verified that the collision manifold in the six-dimensional phase space for (1.20) is three dimensions, and the set of all collision orbits is then four dimensions where time is the fourth dimension. When another mass is included and we have a three-body problem in three dimensions given by (1.2) with $n = 3, \mathbf{x}_k \in \mathbf{R}^3, k = 1, 2, 3$, then the phase space is eighteen dimensional, and for any binary collision, twelve additional free parameters are required. Thus, in this higher dimensional case, the set of all binary collision orbits in the general three-body problem in three dimensions with $\mathbf{c} \neq 0$ is therefore sixteen dimensions.

Thus, in the planar or three-dimensional general three-body problem with $\mathbf{c} \neq \mathbf{0}$, the set of all binary collisions is of smaller dimension than the dimension of the phase space, and hence a set of measure zero. This is summarized in the following lemma.

Lemma 1.16 *The set of all binary collision orbits in the general three-dimensional three-body problem with* $\mathbf{c} \neq 0$ *is a set of measure zero.*

The set of all solutions of (1.39) can be explicitly determined.

For $h < 0$, we write (1.39) as $\mathbf{z}'' + \frac{|h|}{2}\mathbf{z} = 0$. This is just a harmonic oscillator whose general solution is given by

$$\mathbf{z}(s) = \mathbf{a}_1 \cos \lambda s + \mathbf{a}_2 \sin \lambda s, \tag{1.40}$$

$\mathbf{a}_i = (a_{i1}, a_{i2}) \in \mathbf{R}^2, i = 1, 2, \lambda = \sqrt{\frac{|h|}{2}}$.

When $h = 0$, (1.39) reduces to $\mathbf{z}'' = 0$, which yields

$$\mathbf{z}(s) = \mathbf{a}_1 s + \mathbf{a}_2, \tag{1.41}$$

and for $h > 0$,

$$\mathbf{z}(s) = \mathbf{a}_1 e^{\lambda s} + \mathbf{a}_2 e^{-\lambda s}, \tag{1.42}$$

$\lambda = \sqrt{\frac{h}{2}}$.

In section 1.6 the case of general energy is studied from a differential geometric viewpoint.

We conclude this section by considering the case $h < 0$. We show that the position map associated with the Levi-Civita transformation, written in complex coordinates,

$$q = z^2, \tag{1.43}$$

$q \in \mathbb{C}, z \in \mathbb{C}$, maps elliptic motion into elliptic motion. Set $q(\theta) = r(\theta)e^{i\theta}$, where $r = r(\theta)$ is given by (1.27). Equation (1.43) implies

$$z = r^{\frac{1}{2}} e^{i\frac{\theta}{2}},$$

which again represents an elliptical orbit, $z = z(\theta)$. More precisely, in component form, (1.43) is

$$\mathbf{q} = (z_1^2 - z_2^2, 2z_1 z_2). \tag{1.44}$$

We apply this to (1.40), satisfying $z_1(0) = \alpha > 0$, corresponding to the minor axis of an ellipse, where $w_1(0) = z_1'(0) = 0; z_2(0) = 0, w_2(0) = z_2'(0) = \lambda\beta$ (see Figure 1.6).

It is verified that

$$z_1(E) = \alpha \cos \frac{E}{2}, \quad z_2(E) = \beta \sin \frac{E}{2}, \tag{1.45}$$

where $E = 2\lambda s$ is the eccentric anomaly introduced in section 1.3. It is geometrically defined in Figure 1.6. The transformation of the ellipse $\mathbf{z}(E)$ by (1.44) into q_1, q_2 coordinates is shown in Figure 1.4. Equations (1.44), (1.45) imply

$$q_1(E) = -\frac{\beta^2 - \alpha^2}{2} + \frac{\beta^2 + \alpha^2}{2} \cos E, \quad q_2(E) = \alpha\beta \sin E.$$

These imply that the center of the ellipse is located at $q_1 = -\frac{\beta^2 - \alpha^2}{2}, q_2 = 0$. The semimajor axis $a = \frac{\beta^2 + \alpha^2}{2}$. Since the q_1-coordinate of the center of the ellipse, in absolute value, equals ae, then

$$ae = \frac{\beta^2 - \alpha^2}{2}$$

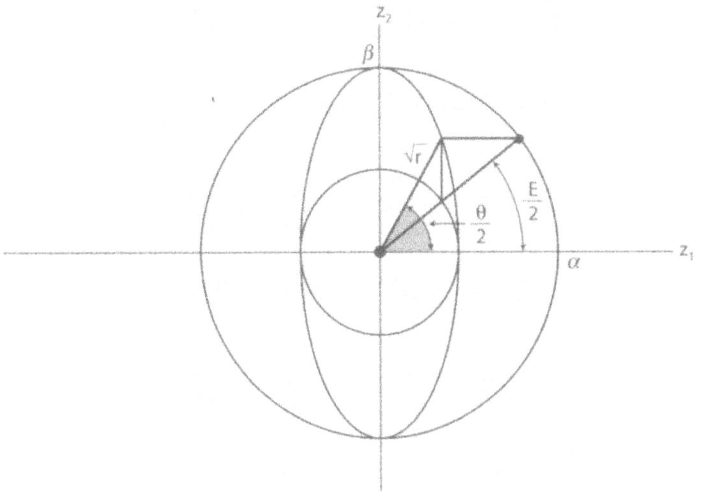

Figure 1.6 Geometry of eccentric anomaly.

and

$$\alpha = \sqrt{a(1-e)}, \beta = \sqrt{a(1+e)}.$$

Then

$$q_1(E) = a(\cos E - e), \quad q_2 = a\sqrt{1-e^2}\sin E. \tag{1.46}$$

Also

$$r(E) = a(1 - e\cos E), \tag{1.47}$$

$\dot{q}_1(E) = -\sqrt{ka}\, r^{-1}\sin E, \dot{q}_2(E) = \sqrt{ka}\, r^{-1}\sqrt{1-e^2}\cos E$. $t(E)$ is obtained from

$$t = \int r(s)ds = \frac{a}{2\lambda}\int (1 - e\cos E)dE$$

or

$$t = \frac{a}{2\lambda}(E - e\sin E), \tag{1.48}$$

where $\lambda = \sqrt{\frac{|h|}{2}} = \frac{1}{2}\sqrt{\frac{k}{a}}$. This yields $\mathcal{M} = E - e\sin E$. (1.48) is called *Kepler's equation*, and $\mathcal{M} = \frac{2\lambda}{a}t = k^{\frac{1}{2}}a^{-\frac{3}{2}}t$, as previously defined at the end of section 1.3.

The Levi-Civita transformation is two-dimensional in position space. Its generalization to three dimensions was initially developed by P. Kustaanheimo [130]. Referring to $q = z^2, q \in \mathbb{C}, z \in \mathbb{C}$ as the Levi-Civita transformation, it maps $\mathbf{R^2} \to \mathbf{R^2}$. The generalization by Kustaanheimo and Stiefel [131], called the *KS transformation*, maps $\mathbf{R^4} \to \mathbf{R^3}$. It is briefly described here (See [214]).

The KS transformation is given by

$$\mathbf{q} = \Lambda(\mathbf{z})\mathbf{z}, \tag{1.49}$$

$$\Lambda(\mathbf{z}) = \begin{pmatrix} z_1 & -z_2 & -z_3 & z_4 \\ z_2 & z_1 & -z_4 & -z_3 \\ z_3 & z_4 & z_1 & z_2 \end{pmatrix},$$

$\mathbf{z} = (z_1, z_2, z_3, z_4) \in \mathbf{R^4}, \mathbf{q} = (q_1, q_2, q_3) \in \mathbf{R^3}$. As with the Levi-Civita transformation, $|\mathbf{q}| = |\mathbf{z}|^2$, and $\mathbf{z} = \mathbf{0} \to \mathbf{q} = \mathbf{0}$, where collision between m_1, m_2 is for the three-dimensional Kepler problem defined by (1.20), with $\mathbf{q} \in \mathbf{R^3}, \mathbf{p} \in \mathbf{R^3}$.

The inverse map to (1.49) is multiple valued, and each point $\mathbf{q}$ is mapped onto a circle S^1 of radius $\sqrt{|\mathbf{q}|}$ on a plane through the origin of $\mathbf{R^4}$. This circle is called a *fiber* in $\mathbf{R^4}$, and fibers corresponding to different points in $\mathbf{R^3}$ do not intersect.

A canonical extension of (1.49) to momentum coordinates $\mathbf{p} \in \mathbf{R^3}$ is obtained as

$$\mathbf{p} = \frac{1}{|\mathbf{z}|^2}\Lambda(\mathbf{z})\mathbf{w}, \tag{1.50}$$

$\mathbf{w} \in \mathbf{R^4}$. Equations (1.49), (1.50) together with $dt = |\mathbf{z}|^2 ds$ are applied to (1.21), where H is again given by (1.19). An analogous system to (1.39) is obtained together with an induced bilinear form, which is an integral:

$$z_4 w_1 - z_3 w_2 + z_2 w_3 - z_1 w_4 = 0.$$

The KS transformation illustrates that an increase in dimension can cause difficulties in generalizing a two-dimensional regularization. This is the case for obtaining a global regularization. However, a local regularization to the three-dimensional case can present no dimension increase, such as Sundman's transformation considered in this section.

An n-dimensional regularization of the n-dimensional Kepler problem is described in section 1.6; it avoids the difficulties from which the KS transformation suffers of being multiple valued. This regularization is closely tied in with the basic Riemannian geometries of constant curvature.

1.5 THE RESTRICTED THREE-BODY PROBLEM: FORMULATIONS

Throughout most of this book, we will be interested in the three-body problem under special restrictions. This version of the three-body problem is derived in different coordinate systems and under various assumptions.

Consider the planar three-body problem in Jacobi coordinates (1.12), (1.13) for $\mathbf{q}(t), \mathbf{Q}(t)$. Recall that $\mathbf{q}(t)$ describes the motion of P_2 about P_1, and $\mathbf{Q}(t)$ describes the motion of P_3 relative to the center of mass of P_1, P_2, located at $\boldsymbol{\beta} = \nu^{-1}(m_1\mathbf{x}_1 + m_2\mathbf{x}_2)$, as shown in Figure 1.1 in an inertial coordinate system. The origin of this coordinate system is the center of mass $\boldsymbol{\rho}$ of the three mass particles.

We will regard P_1, P_2 as planetary objects, such as the Sun and Jupiter, respectively, or the Earth and Moon, which move in elliptical orbits. P_2 will be regarded as the smaller of P_1, P_2. In our two examples of P_1, P_2, the mass ratio $\mu = m_2/(m_1 + m_2)$ is significantly smaller than 1. For Jupiter and the Sun, $\mu = 0.001$, and for Moon and Earth, $\mu = 0.012$. If P_1 = Earth and P_2 = Sun, then $\mu = 0.000003$. Thus $\mu \ll 1$. More generally, we assume, unless otherwise notified, that $0 \le \mu < 1/2$. The assumption on P_3 is that it has negligible mass relative to P_1, P_2. For example, P_3 could be a small asteroid, comet, or spacecraft. Because P_1, P_2 are planetary-sized objects the gravitational perturbation on them due to P_3 will be negligible. Thus, P_1, P_2 can be viewed as a decoupled binary system, and their relative motion will reduce to the Kepler two-body problem (1.20). This is seen from (1.12) if we set $m_3 = 0$. By our assumptions, $m_3 \approx 0$. On the other hand, the motion of the particle P_3 of zero mass will be gravitationally perturbed by the two-body elliptical motion of P_1, P_2. The motion of P_3 is defined by (1.13), which yields a well-defined system as $m_3 \to 0$.

Setting $m_3 = 0$ implies $M = \nu$, and also

$$\boldsymbol{\beta} = \boldsymbol{\rho} = 0. \tag{1.51}$$

Thus, the origin of the coordinate system corresponds to the center of mass of P_1, P_2. Another restriction we make is to perform a time scaling so that $G = 1$. This scaling is given by $t \to \sqrt{G}t = \tilde{t}$, i.e., $d/dt = \sqrt{G}d/d\tilde{t}$. Thus, with $G = 1$, $\tilde{t}$ is the new time variable, and we relabel t for notational convenience. We also scale the masses m_1, m_2 so that $\nu = m_1 + m_2 = 1$. The mass is therefore dimensionless. With this normalization we set $m_1 = 1 - \mu, m_2 = \mu$, where $\mu = m_2/(m_1 + m_2)$.

As a final restriction, we assume that P_1, P_2 describe circular orbits, i.e., $e = 0$ in (1.28), implying $r = a = k/2|h|$ by (1.29), where $k = G\nu = 1$. Choosing the unit of length so that $r = |\mathbf{q}| = a = 1$ (i.e., $H = h = -\frac{1}{2}$) defines the *planar circular restricted three-body problem*, or *restricted problem* for short. Thus, it is defined in inertial coordinates by

$$\ddot{\mathbf{Q}} = -\frac{1-\mu}{r_{13}^3}(\mathbf{Q} - m_2\mathbf{q}(t)) + \frac{\mu}{r_{23}^3}(\mathbf{Q} - m_1\mathbf{q}(t)), \tag{1.52}$$

where $\mathbf{q}(t)$ defines the circular motion of P_2 about P_1 in an inertial coordinate system centered at P_1. The center of mass $\boldsymbol{\beta}$ of P_1, P_2 is at the origin, and according to (1.46),

$$q_1(t) = \cos t, \quad q_2(t) = \sin t. \tag{1.53}$$

Equation (1.48) implies $t = E = s$. The angular velocity of P_2 about P_1 is 1, and the period of motion is 2π. The relative distances of P_3 to P_1 and P_2 are

$$r_{13}(t) = \sqrt{(Q_1 + \mu C)^2 + (Q_2 + \mu S)^2},$$
$$r_{23}(t) = \sqrt{(Q_1 - (1-\mu)C)^2 + (Q_2 - (1-\mu)S)^2} \tag{1.54}$$

where $C \equiv \cos t, S \equiv \sin t$.

Definition 1.17 *Equation (1.52) defines the restricted problem in* inertial coordinates *whose origin is the center of mass of* P_1, P_2, *where* $m_1 = 1 - \mu, m_2 = \mu, 0 \leq \mu < 1/2, r = |\mathbf{q}| = 1$.

Equation (1.52) can be written as

$$\ddot{\mathbf{Q}} = \Omega_{\mathbf{Q}}, \tag{1.55}$$

where

$$\Omega = \frac{1-\mu}{r_{13}} + \frac{\mu}{r_{23}}. \tag{1.56}$$

Ω is the potential energy of m_3, $\Omega = \Omega(\mathbf{Q}, t)$. For $\mu = 0$, the mass of P_2 vanishes, and (1.55) reduces to the two-body problem in inertial coordinates between P_3 and P_1 of mass $m_1 = 1$ at the origin. In this case (1.53) reduces to (1.20) with $G = \nu = 1$, with $\mathbf{q}$ replaced by $\mathbf{Q}$, and $r_{13} = |\mathbf{Q}|$.

Lemma 1.18 *The total energy* $\mathcal{H}$ *of the full system for* P_1, P_2, P_3 *for the restricted problem is*

$$\mathcal{H} = -\frac{m_1 m_2}{2}. \tag{1.57}$$

Proof. For the planar three-body problem, $\mathcal{H}$ is given by (1.16). Following the sequence of normalizations made for the restricted problem we set $m_3 = 0$, which implies $k_1 = 0, \mathbf{P} = \mathbf{0}, k_2 = m_1 m_2 \nu^{-1}$. With the normalization $G = 1, m_1 + m_2 = 1$, (1.16) reduces to

$$\mathcal{H} = \frac{1}{2} m_1 m_2 |\dot{\mathbf{q}}|^2 - \frac{m_1 m_2}{|\mathbf{q}|}.$$

Therefore by (1.29), where $H = h = -1/2$,

$$\mathcal{H} = m_1 m_2 \left[\frac{1}{2}|\dot{\mathbf{q}}|^2 - \frac{1}{|\mathbf{q}|}\right] = m_1 m_2 h = -\frac{m_1 m_2}{2}. \qquad \square$$

The restricted problem can be described as determining the motion of a particle of zero mass in a gravitational field generated by the uniform circular motion of the mass points P_1, P_2, called the *primaries*. Since $m_3 = 0$, (1.57) can be viewed as the energy of P_1, P_2. It is an integral for the two-body motion of P_1, P_2.

The restricted problem possesses an integral associated to the energy of only P_3.

Definition 1.19 *The restricted problem (1.55) has an energy integral called the* Jacobi integral, *and in* inertial coordinates *it is given by*

$$J = J(\mathbf{Q}, \dot{\mathbf{Q}}, t) = -|\dot{\mathbf{Q}}|^2 + 2(Q_1\dot{Q}_2 - Q_2\dot{Q}_1) + 2\Omega. \qquad (1.58)$$

The term $c = Q_1\dot{Q}_2 - Q_2\dot{Q}_1$ is the *angular momentum of* m_3, which in polar coordinates r, θ is equivalent to $c = r^2\dot{\theta}, r = \sqrt{Q_1^2 + Q_2^2}, \theta = \arctan(\frac{Q_2}{Q_1})$. J is alternatively referred to as the *Jacobi energy.*

Lemma 1.20 *J is an integral of (1.55).*

Proof. Let $\mathbf{Q} = \mathbf{Q}(t)$ represent a solution of (1.55). We need to show

$$\frac{d}{dt}J(\mathbf{Q}, \dot{\mathbf{Q}}, t) = 0.$$

Now,

$$\begin{aligned}\frac{d}{dt}J &= (J_{\mathbf{Q}}, \dot{\mathbf{Q}}) + (J_{\dot{\mathbf{Q}}}, \ddot{\mathbf{Q}}) + J_t \\ &= 2(\Omega_{\mathbf{Q}}, \dot{\mathbf{Q}}) - 2(\dot{\mathbf{Q}}, \ddot{\mathbf{Q}}) + 2(Q_1\ddot{Q}_2 - Q_2\ddot{Q}_1) + 2\Omega_t.\end{aligned}$$

By (1.55) the first two terms cancel, and

$$\frac{1}{2}\frac{d}{dt}J = Q_1\Omega_{Q_2} - Q_2\Omega_{Q_1} + \Omega_t.$$

We now verify that $Q_1\Omega_{Q_2} - Q_2\Omega_{Q_1} = -\Omega_t$, thus yielding the proof.

Using (1.56), direct calculation yields

$$\begin{aligned}&Q_1\Omega_{Q_2} - Q_2\Omega_{Q_1} \\ &= -\frac{\mu(1-\mu)}{r_{13}^3}(Q_1S - Q_2C) - \frac{\mu(1-\mu)}{r_{23}^3}(-Q_1S + Q_2C)\end{aligned}$$

and

$$\begin{aligned}\Omega_t &= -\frac{1-\mu}{r_{13}^3}\left[(Q_1 + \mu C)(-\mu S) + (Q_2 + \mu S)\mu C\right] \\ &\quad -\frac{\mu}{r_{23}^3}\left[(Q_1 - (1-\mu)C)(1-\mu)S - (Q_2 - (1-\mu)S)(1-\mu)C\right] \\ &= -\left[Q_1\Omega_{Q_2} - Q_2\Omega_{Q_1}\right].\end{aligned}$$

Therefore, $\dot{J} = 0$. □

Lemma 1.20 implies the following lemma.

Lemma 1.21 *The set*

$$J^{-1}(C) = \left\{(\mathbf{Q}, \dot{\mathbf{Q}}, t) \in \mathbf{R}^5 | J = C, C \in \mathbf{R}^1\right\}$$

is a four-dimensional surface in the five-dimensional extended phase space on which the solution $\mathbf{Q}(t), \dot{\mathbf{Q}}(t)$ *lies for a given value of* C. *(See Figure 1.7.) (The constant* C *should not be confused with* $C = \cos t$*)*

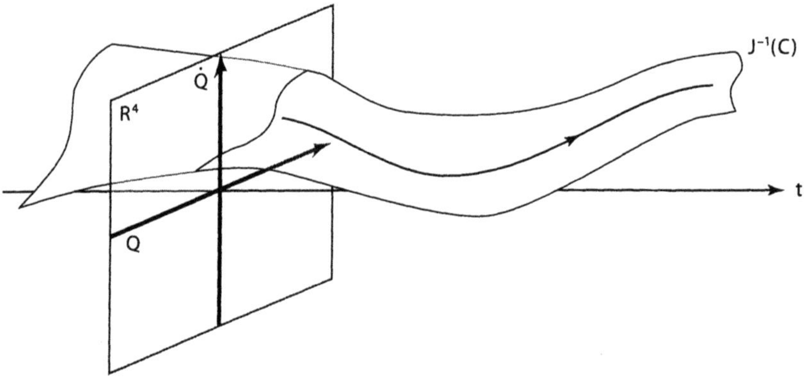

Figure 1.7 Jacobi integral surface.

Definition 1.22 C *is called the* Jacobi constant, *or* Jacobi energy.

Equation (1.55) can be transformed to a new coordinate system which is *time independent.* That is, it is an *autonomous system.* Since P_1, P_2 move about the origin with constant angular velocity of 1, then transforming to a new coordinate system x_1, x_2, which also rotates with an angular velocity ω of 1, implies that P_1, P_2 will be fixed. (See Figure 1.8.)

Without loss of generality, we assume that P_1, P_2 lie fixed on the x_1-axis.

Definition 1.23 *The* x_1, x_2*-coordinate system is called a* rotating coordinate system *or* fixed coordinate system.

The transformation $\mathbf{Q} \to \mathbf{x}$ is thus a uniform rotation given by an orthogonal matrix $\mathbf{R}(t)$,

$$\mathbf{x} = \mathbf{R}(t)\mathbf{Q}, \tag{1.59}$$

$$\mathbf{R}(t) = \begin{pmatrix} \cos t & \sin t \\ -\sin t & \cos t \end{pmatrix}.$$

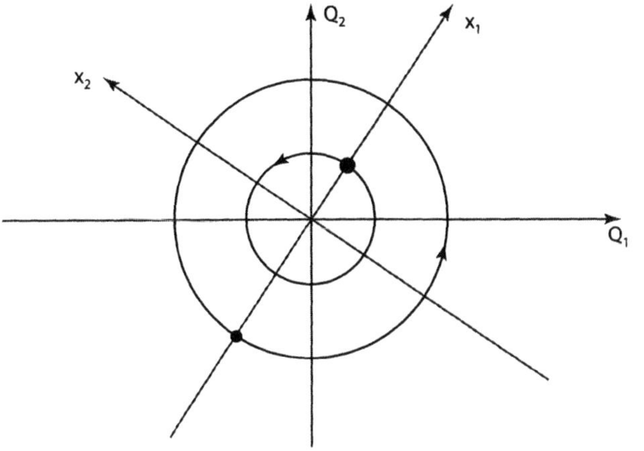

Figure 1.8 Rotating and inertial coordinates.

The inverse transformation is

$$\mathbf{Q} = \mathbf{R}^{-1}(t)\mathbf{x}, \tag{1.60}$$

where

$$R^{-1}(t) = \begin{pmatrix} \cos t & -\sin t \\ \sin t & \cos t \end{pmatrix}.$$

It is verified that substitution of (1.60) (i.e., $Q_1 = x_1 \cos t - x_2 \sin t, Q_2 = x_1 \sin t + x_2 \cos t$), into (1.55) yields the restricted problem in *rotating or fixed* coordinates,

$$\begin{aligned} \ddot{x}_1 - 2\dot{x}_2 &= x_1 + \tilde{\Omega}_{x_1}, \\ \ddot{x}_2 + 2\dot{x}_1 &= x_2 + \tilde{\Omega}_{x_2}, \end{aligned} \tag{1.61}$$

where

$$\tilde{\Omega} = \tilde{\Omega}(\tilde{x}) = \Omega(\mathbf{R}^{-1}\mathbf{x}) = \frac{1-\mu}{r_{13}} + \frac{\mu}{r_{23}}, \tag{1.62}$$

$$r_{13} = \sqrt{(x_1 - \mu)^2 + x_2^2}, \quad r_{23} = \sqrt{(x_1 - (-1+\mu))^2 + x_2^2}.$$

The particle P_1 is fixed at $\mathbf{x} = (\mu, 0)$, and P_2 is fixed at $\mathbf{x}_2 = (-1+\mu, 0)$. (See Figure 1.9.) The system (1.61) is a standard form of the restricted problem commonly seen in the literature.

The system (1.61) is an autonomous system of differential equations.

When $\mu = 0$, (1.61) reduces to the two-body problem in rotating coordinates between P_3 of zero mass and P_1 at the origin of unit mass, and $r_{13} = |\mathbf{x}| = |\mathbf{Q}|$. The system (1.62) has the same form as (1.56), except t is not present, and r_{13}, r_{23} are simplified.

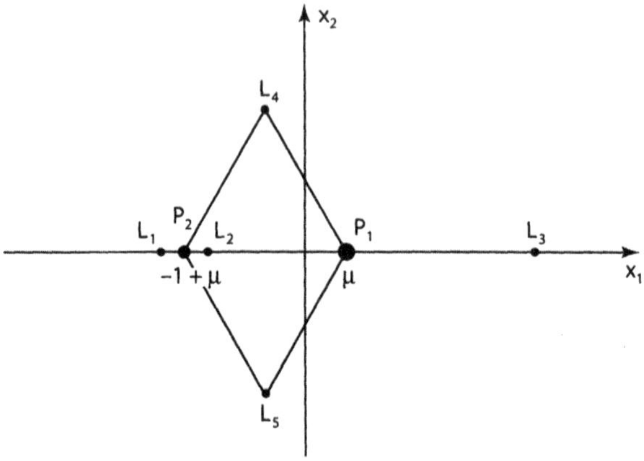

Figure 1.9 Rotating coordinate system showing Lagrange points.

The right-hand side of (1.61) represents the sum of two forces $\mathbf{F}, \mathbf{G}$,

$$\mathbf{F} = (x_1, x_2), \qquad \mathbf{G} = (\tilde{\Omega}_{x_1}, \tilde{\Omega}_{x_2}).$$

$\mathbf{F}$ is the outward radially directed centrifugal force, and $\mathbf{G}$ is the sum of the gravitational forces due to m_1, m_2 on m_3. $\mathbf{F}$ exists due to the fact of being in a rotating coordinate system and is an artificially induced force.

By direct substitution of (1.60) into (1.58) we obtain the Jacobi integral in rotating coordinates. It is verified that

$$\tilde{J}(\mathbf{x}, \dot{\mathbf{x}}) = J(\mathbf{R}^{-1}(t)\mathbf{x}, \quad \frac{d}{dt}\mathbf{R}^{-1}(t)\mathbf{x}) = -|\dot{\mathbf{x}}|^2 + r^2 + 2\tilde{\Omega}(\mathbf{x}), \tag{1.63}$$

where $r = |\mathbf{x}| = |\mathbf{R}(t)\mathbf{Q}| = |\mathbf{Q}|$. It is verified by direct differentiation of (1.63) using (1.61) that indeed $\frac{d}{dt}\tilde{J} = 0$.

Definition 1.24 $\tilde{J}(\mathbf{x}, \dot{\mathbf{x}})$ *is called the Jacobi integral in* rotating coordinates.

Lemma 1.25 *The set*

$$\tilde{J}^{-1}(C) = \{(\mathbf{x}, \dot{\mathbf{x}}) \in \mathbf{R}^4 | \tilde{J} = C\} \tag{1.64}$$

is a three-dimensional surface in the four-dimensional phase space on which the solutions $x(t), \dot{x}(t)$ of (1.61) lie for a given value of C.

Thus, the dimension of the Jacobi integral manifold for the given value of C has decreased by 1 when changing from the Q_1, Q_2 coordinates to the x_1, x_2 coordinates.

The Jacobi integral (1.58) for (1.55) can be written in a more standardized form,

$$\hat{J} = \frac{1}{2}|\dot{\mathbf{Q}}|^2 - (Q_1\dot{Q}_2 - Q_2\dot{Q}_1) - \Omega(\mathbf{Q}, t) \tag{1.65}$$

or

$$\hat{J} = E - c, \tag{1.66}$$

where

$$E = \frac{1}{2}|\dot{\mathbf{Q}}|^2 - \Omega(\mathbf{Q}, t) \tag{1.67}$$

is the total energy of P_3 in the barycentric inertial system (i.e., where the center of mass is at the origin) and

$$c = Q_1\dot{Q}_2 - Q_2\dot{Q}_1 \tag{1.68}$$

is the angular momentum. For $\mu \neq 0$ neither E nor c are integrals. They are both integrals when $\mu = 0$, where $m_2 = 0$ and E reduces to the two-body problem between P_1 of mass $m_1 = 1$ and P_3. In that case, E reduces to (1.19).

It is verified that

$$L(\mathbf{Q}, \dot{\mathbf{Q}}) = L(\mathbf{x}, \dot{\mathbf{x}}) + |\mathbf{x}|^2, \tag{1.69}$$

where $L(\mathbf{x}, \dot{\mathbf{x}}) = \dot{x}_1 x_2 - \dot{x}_2 x_1$. $L(\mathbf{Q}, \dot{\mathbf{Q}})$ is an integral for the two-body problem (1.18) with $\mathbf{q}$ replaced by $\mathbf{Q}$. This problem has energy (1.19), which we more generally write as

$$H = \frac{1}{2}|\dot{\mathbf{Q}}|^2 - \frac{a}{|\mathbf{Q}|},$$

where $a \in \mathbf{R}^1$ is a constant. Since $L(\mathbf{Q}, \dot{\mathbf{Q}})$ is an integral for the Kepler flow defined by (1.18) in inertial coordinates, then $L(\mathbf{x}, \dot{\mathbf{x}}) + |\mathbf{x}|^2$ is an integral for the Kepler flow in rotating coordinates, given by (1.61) with $\mu = 0$, where $r_{13} = |\mathbf{x}|$.

It is also verified by (1.60) that (1.19), with $G\nu$ replaced by a, in rotating coordinates is

$$\tilde{H}(\mathbf{x}, \dot{\mathbf{x}}) = H(R^{-1}\mathbf{x}, \frac{d}{dt}(R^{-1}\mathbf{x})) = \frac{1}{2}|\dot{\mathbf{x}}|^2 + \frac{1}{2}|\mathbf{x}|^2 - L(\mathbf{x}, \dot{\mathbf{x}}) - \frac{a}{|\mathbf{x}|}, \tag{1.70}$$

which is therefore an integral of the Kepler flow in rotating coordinates. Thus, in summary, we have the following lemma.

Lemma 1.26

$$\tilde{L} = L(\mathbf{x}, \dot{\mathbf{x}}) + |\mathbf{x}|^2,$$
$$\tilde{H} = \frac{1}{2}|\dot{\mathbf{x}}|^2 + \frac{1}{2}|\mathbf{x}|^2 - L(\dot{\mathbf{x}}, \dot{\mathbf{x}}) - \frac{a}{|\mathbf{x}|},$$

corresponding to the angular momentum and energy of the two-body problem in rotating coordinates, are integrals of the Kepler flow in rotating coordinates,

$$\ddot{x}_1 - 2\dot{x}_2 = x_1 - \frac{a}{|\mathbf{x}|^3},$$
$$\ddot{x}_2 - 2\dot{x}_1 = x_2 - \frac{a}{|\mathbf{x}|^3}.$$

It is noted that

$$\tilde{J}|_{\mu=0} = -2\left[\tilde{H} - \tilde{L}\right], \tag{1.71}$$

where $a = 1$. More generally we calculate $\tilde{J}$ for $\mu \neq 0$ in terms of $\tilde{H}$ in P_2-centered coordinates in chapter 3, section 3.2.

Equation (1.55) can be written in Hamiltonian form:

$$\dot{\mathbf{Q}} = E_{\mathbf{P}},\ \dot{\mathbf{P}} = -E_{\mathbf{Q}}, \tag{1.72}$$

where E is given by (1.67) with $\dot{\mathbf{Q}} = \mathbf{P}$. E is not an integral for $\mu \neq 0$ because it is time dependent: $\dot{E} = (\mathbf{P}, \dot{\mathbf{P}}) - (\Omega_{\mathbf{Q}}, \dot{\mathbf{Q}}) - \Omega_t = -\Omega_t \neq 0$.

Likewise, for rotating coordinates, (1.61) can be written as a Hamiltonian system,

$$\dot{\mathbf{q}} = G_{\mathbf{p}}, \quad \dot{\mathbf{p}} = -G_{\mathbf{q}}, \tag{1.73}$$

where

$$G = \frac{1}{2}|\mathbf{p}|^2 - (q_1 p_2 - q_2 p_1) - \tilde{\Omega}(\mathbf{q}) \tag{1.74}$$

and G is an integral for (1.73) since it is time independent.

Lemma 1.27 *Equation (1.61) is obtained from (1.73) by the map*

$$q_k = x_k, \quad p_1 = \dot{x}_1 - x_2, \quad p_2 = \dot{x}_2 + x_1, \tag{1.75}$$

and

$$G(\mathbf{p}, \mathbf{q}) = -\frac{1}{2}\tilde{J}(\mathbf{x}, \dot{\mathbf{x}}).$$

Proof.

$$\dot{\mathbf{q}} = \mathbf{p} + \begin{pmatrix} q_2 \\ -q_1 \end{pmatrix},$$
$$\dot{\mathbf{p}} = \begin{pmatrix} p_2 \\ -p_1 \end{pmatrix} - \tilde{\Omega}_{\mathbf{q}}(\mathbf{q}).$$

Therefore, these yield, respectively,

$$\dot{\mathbf{x}} = \begin{pmatrix} \dot{x}_1 - x_2 \\ \dot{x}_2 + x_1 \end{pmatrix} + \begin{pmatrix} x_2 \\ -x_1 \end{pmatrix}$$

and

$$\begin{pmatrix} \ddot{x}_1 - \dot{x}_2 \\ \ddot{x}_2 + \dot{x}_1 \end{pmatrix} = \begin{pmatrix} \dot{x}_2 + x_1 \\ -\dot{x}_1 + x_2 \end{pmatrix} - \tilde{\Omega}_{\mathbf{x}}(\mathbf{x}),$$

which yields (1.61).

$$\begin{aligned} &G(\mathbf{p}(\mathbf{x}, \dot{\mathbf{x}}), \mathbf{q}(\mathbf{x}, \dot{\mathbf{x}})) = \tilde{G}(\mathbf{x}, \dot{\mathbf{x}}) \\ &= \frac{1}{2}[(\dot{x}_1 - x_2)^2 + (\dot{x}_2 + x_1)^2] - [x_1(\dot{x}_2 + x_1) - x_2(\dot{x}_1 - x_2)] - \tilde{\Omega}(\mathbf{x}) \\ &= \frac{1}{2}|\dot{\mathbf{x}}|^2 - \dot{x}_1 x_2 + \dot{x}_2 x_1 + \frac{1}{2}|\mathbf{x}|^2 - x_1\dot{x}_2 + x_2\dot{x}_1 - |x|^2 - \tilde{\Omega}(\mathbf{x}). \end{aligned}$$

Thus,

$$\tilde{G} = \frac{1}{2}|\dot{\mathbf{x}}|^2 - \frac{1}{2}r^2 - \tilde{\Omega}(\mathbf{x}) = -\frac{1}{2}\tilde{J}. \tag{1.76}$$

□

Note that although G is a Hamiltonian function for (1.73), neither $\tilde{G}$ nor $\tilde{J}$ are Hamiltonian functions for (1.61), but they are integrals for that system. Equation (1.61) is not in canonical form since (1.75) is not a canonical map. This follows since

$$\begin{aligned} (\mathbf{dp}, \mathbf{dq}) &= (dy_1 - dx_2)dx_1 + (dy_2 + dx_1)dx_2 \\ &= (\mathbf{dy}, \mathbf{dx}) + 2dx_1dx_2, \end{aligned}$$

where we used the antisymmetry of the products of one-forms dx, dy $dxdy = -dydx$ (see [15]). Thus $(\mathbf{dp}, \mathbf{dq}) \neq (\mathbf{dy}, \mathbf{dx})$.

Definition 1.28 *The* three-dimensional restricted problem *in* inertial coordinates *is given by (1.55), (1.56) with* $\mathbf{Q} = (Q_1, Q_2, Q_3) \in \mathbf{R^3}$, *and*

$$\begin{aligned} r_{13} &= \sqrt{(Q_1 + \mu\cos t)^2 + (Q_2 + \mu\sin t)^2 + Q_3^2}, \\ r_{23} &= \sqrt{(Q_1 - (1-\mu)\cos t)^2 + (Q_2 - (1-\mu)\sin t)^2 + Q_3^2}, \end{aligned}$$

where J is given by (1.58). The transformation to rotating coordinates $\mathbf{x} = (x_1, x_2, x_3) \in \mathbf{R^3}$ *is given by*

$$\mathbf{x} = S(t)\mathbf{Q},$$

where

$$S(t) = \begin{pmatrix} \mathbf{R}(t) & 0 \\ 0 & 1 \end{pmatrix},$$

leaving the Q_3-axis invariant. This maps (1.55) into the system given by (1.61) together with $\ddot{x}_3 = \tilde{\Omega}_{x_3}$, where $r_{13} = \sqrt{(x_1 - \mu)^2 + x_2^2 + x_3^2}, r_{23} = \sqrt{(x_1 - (-1+\mu))^2 + x_2^2 + x_3^2}$. The Jacobi integral is given by (1.63), where $r = |\mathbf{x}| = |\mathbf{S}(t)\mathbf{Q}|$.

A final definition for this section is for the *three-dimensional elliptic restricted three-body problem* in inertial coordinates. It parallels the derivation of (1.52), (1.55). We begin with (1.11) with $\mathbf{x}_k \in \mathbf{R^3}$, and obtain (1.12), (1.13), where $\mathbf{q} \in \mathbf{R^3}, \mathbf{Q} \in \mathbf{R^3}$. This represents the three-dimensional three-body problem in Jacobi coordinates. Setting $m_3 = 0$, the center of mass of P_1, P_2 is at the origin, expressed in (1.51); the time is scaled so that $G = 1$; and $m_1 + m_2 = 1, m_1 = 1 - \mu, m_2 = \mu, 0 \leq \mu < 1/2$. Instead of choosing $e = 0$ in (1.28), it is more generally assumed that $e \in [0, 1)$, where r is given by (1.28) for $h < 0$, expressed as $r = r(\theta), \theta \in [0, 2\pi]$. The unit of length is normalized so that the semimajor axis $a = 1$, or equivalently $h = -\frac{1}{2}$.

For the binary system m_1, m_2, since the motion is planar it is assumed without loss of generality to lie in the q_1, q_2-plane so that for the elliptic motion of this pair, $q_3 = 0$. Thus, in (1.12) $\mathbf{q} = \mathbf{q}(t) = (q_1(t), q_2(t), 0)$, where $q_1(t), q_2(t)$ are defined by (1.18), with $G\nu = 1$.

Equation (1.55) is again obtained with two changes. First, $\mathbf{Q} \in \mathbf{R^3}$ and r_{13}, r_{23} take a different form.

Definition 1.29 *The three-dimensional elliptic restricted three-body problem is defined by*

$$\ddot{\mathbf{Q}} = \Omega_{\mathbf{Q}}, \tag{1.77}$$

$\mathbf{Q} \in \mathbf{R^3}$, *with,* $\boldsymbol{\Omega}$ *given by (1.56), where*

$$r_{13} = |\mathbf{Q} + \mu\mathbf{q}(t)|, r_{23} = |\mathbf{Q} - (1 - \mu)\mathbf{q}(t)|; \tag{1.78}$$

$\mathbf{q}(t) = (q_1(t), q_2(t), 0), q_k(t), k = 1, 2,$ *are given by (1.46) with* $a = 1$ *and* E *is given implicitly by (1.48).*

Equation (1.77) reduces to the three-dimensional restricted problem when $e = 0$. Thus, the three-dimensional elliptic restricted problem in Definition 1.29 describes the motion of a zero mass particle P_3 in a Newtonian gravitational field generated by the Keplerian elliptic motion of the binary pair P_1, P_2.

If $0 < \mu \ll 1$ P_1 lies very close to the origin and P_2 moves about the center of mass of P_1, P_2, which lies very close to P_1, then P_2 moves approximately about P_1, approximately at the origin. If, moreover, $e \gtrsim 0$, then the situation approximates the motions of most of the planets of our solar system moving about the Sun. Notable exceptions are Pluto and Mercury, where $e = 0.247, 0.206$, respectively. Also, evidence is pointing to the fact that Pluto is not a planet but rather belongs to a different class of objects called Kuiper belt objects [41, 21]. For all the other planets $e \doteq 0.0n, n \in \{0, 1, \ldots, 9\}$. Also note that the planetary ephemeris which accurately describes the motions of the planets [212], e, as well as the other orbital parameters, are not actually constant, and have small variations. This is due primarily to the fact that the Sun and planets are not mass points and are spherical in shape, with finite radius. Also, they are not purely spherical and have oblateness perturbations which distort the ideal gravitational field. Moreover, all the planets gravitationally interact with one another, and their orbits are gravitationally perturbed by neighboring stars. In addition, our solar system itself orbits the center mass of the Milky Way galaxy. There are also non-gravitational perturbations, and forces not understood or known.

Nevertheless, the variations in the orbital parameters of the planets are negligible and the ideal model we are using by viewing the Sun and planets as point masses yields accurate results when compared to reality in many situations.

The existence of solutions to the planar restricted problem in rotating coordinates (1.61) is addressed in chapter 2, where quasi-periodic motion is proven to exist by the Kolmogorov-Arnold-Moser(KAM) theorem. Reduction of the flow of the restricted problem to a monotone twist map proves the existence of chaotic motion due to the complicated intersection of hyperbolic invariant manifolds, and also by applying Aubrey-Mather theory.

In chapter 3, different types of capture solutions for the restricted problem are proven to exist and parabolic orbits are studied. Some of the capture solutions lie near parabolic orbits and are proven to be associated to a hyperbolic invariant set which gives rise to chaotic motion. Applications are also discussed.

1.6 THE KEPLER PROBLEM AND EQUIVALENT GEODESIC FLOWS

In this section we prove that the flow of the Kepler problem is equivalent to the geodesic flow of the basic Riemannian spaces of constant Gaussian curvature K. These spaces turn out to be topologically equivalent to the fixed energy surfaces of the Kepler problem. This is done for $n \geq 2$ dimensions and provides an n-dimensional regularization of collision in the Kepler problem. The regularized differential equations are just the differential equations for the geodesic flows and are easily solved explicitly. The proof of this result follows from Moser [174], Osipov [182, 183], and Belbruno [22, 24].

We will describe the basic results for the three Kepler energies, $H < 0, H > 0, H = 0$, where H is the Kepler energy. These results will be described separately for each respective energy case, and the basic theorems stated. We will then give a new proof for all the cases at once. The n-dimensional Kepler problem is defined by

$$\ddot{\mathbf{q}} = -\frac{\mathbf{q}}{|\mathbf{q}|^3}, \tag{1.79}$$

$\cdot \equiv \frac{d}{dt}, \mathbf{q} = (q_1, q_2, \ldots, q_n) \in \mathbf{R^n}$, which can be written as a Hamiltonian system

$$\dot{\mathbf{q}} = H_{\mathbf{p}}, \quad \dot{\mathbf{p}} = -H_{\mathbf{q}}, \tag{1.80}$$

$\mathbf{p} = (p_1, p_2, \ldots, p_n) \in \mathbf{R^n}, n \geq 2,$

$$H = \frac{1}{2}|\mathbf{p}|^2 - \frac{1}{|\mathbf{q}|}. \tag{1.81}$$

Equations (1.79), (1.80) are just a generalization of (1.20), (1.21) to n dimensions. We define the $(2n-1)$-dimensional energy surface

$$H^{-1}(h) = \left\{(\mathbf{p}, \mathbf{q}) \in \mathbf{R^{2n}} | H = h, h \in \mathbf{R^1}\right\}. \tag{1.82}$$

We consider the three basic cases, $h = -\frac{1}{2}, \frac{1}{2}, 0$. The case of general $h < 0, h > 0$ follows by a simple scaling of $h = \frac{1}{2}, \frac{1}{2}$, respectively.

1.6.1 Case $h = -\frac{1}{2}$

In this case, the Kepler problem turns out to be topologically equivalent with the geodesic flow on the unit sphere, $S^n = \left\{\boldsymbol{\xi} = (\xi_0, \xi_1, \ldots, \xi_n) \in \mathbf{R^{n+1}} | |\boldsymbol{\xi}|^2 = \sum_{k=0}^n \xi_k^2 = 1\right\}$. The equivalence is accomplished by the stereographic projection

$$p_k = \frac{\xi_k}{1 - \xi_0}, \quad k = 1, 2, \ldots, n, \tag{1.83}$$

$\xi_0 \neq 1$. See Figure 1.10.

This maps the great circles on $S_0^n = S^n \backslash \{\xi_0 = 1\}$, which are the geodesics, onto circles in $\mathbf{p}$-space. $\{\xi_0 = 1\} \equiv \{\xi_0 = 1, \xi_1 = \xi_2 = \cdots = \xi_n = 0\}$ is the north pole of the sphere. The notation $S^n \backslash \{\xi_0 = 1\}$ means that the north pole point is deleted from the sphere. The circles in $\mathbf{p}$-space are called *hodographs* and correspond to the paths traced out by the velocity components of an elliptical path in $\mathbf{q}$-space. This type of equivalence first goes back to Fock [83], who applied it to the momentum variables to transform the Schrödinger wave equation in quantum mechanics. This was done for the case of S^3. The use of the projection (1.83) for Kepler's problem, also for $n = 3$, goes back to Györgyi [90], who canonically extended the map (1.83) to a mapping $q_k = q_k(\boldsymbol{\xi}, \boldsymbol{\eta}), \boldsymbol{\eta} \in \mathbf{R^4}$, where $\boldsymbol{\eta} = (\eta_0, \eta_1, \eta_2, \eta_3)$ is the velocity vector of a geodesic on S^3, therefore normal to $\boldsymbol{\xi}$. A general equivalence of the Kepler problem with the geodesic flow on $S^n, n \geq 2$, for $h < 0$ was done by Moser [174].

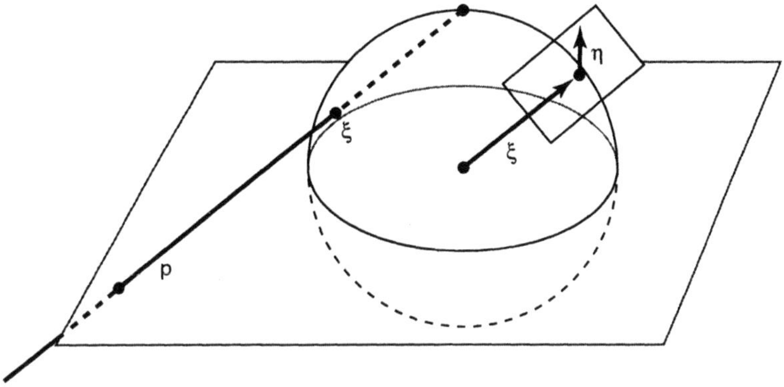

Figure 1.10 Stereographic projection from S^2.

The canonical extension of (1.83) in [174] is found by satisfying the relationship

$$\langle \boldsymbol{\eta}, \mathbf{d}\boldsymbol{\xi} \rangle = \langle \mathbf{y}, \mathbf{dx} \rangle, \tag{1.84}$$

$\mathbf{x} \in \mathbf{R^n}, \mathbf{y} \in \mathbf{R^n}, \boldsymbol{\eta} = (\eta_0, \eta_1, \ldots, \eta_n) \in \mathbf{R^{n+1}}$ and then setting $\mathbf{y} = \mathbf{q}, \mathbf{x} = -\mathbf{p}, \mathbf{x} = (x_1, \ldots, x_n) \in \mathbf{R^n}, \mathbf{y} = (y_1, \ldots, y_n) \in \mathbf{R^n}, \langle \mathbf{y}, \mathbf{dx} \rangle \equiv \sum_{k=1}^n y_k dx_k$, $\langle \boldsymbol{\eta}, \mathbf{d}\boldsymbol{\xi} \rangle \equiv \sum_{k=0}^n \eta_k d\xi_k$. The variables $(\boldsymbol{\xi}, \boldsymbol{\eta}) \in T_1(S^n)$,

$$T_1(S^n) = \left\{ (\boldsymbol{\xi}, \boldsymbol{\eta}) \in \mathbf{R^{2n+2}} | |\boldsymbol{\xi}| = 1, |\boldsymbol{\eta}| = 1, \langle \boldsymbol{\xi}, \boldsymbol{\eta} \rangle = 0 \right\}, \tag{1.85}$$

where $|\boldsymbol{\xi}|^2 = \langle \boldsymbol{\xi}, \boldsymbol{\xi} \rangle$. $T_1(S^n)$ is called the *unit tangent bundle of* S^n. This means that all tangent vectors to S^n have unit length. *Tangent bundle* means the collection of all vectors in the tangent space of S^n.

It is found after simplification that

$$q_k = \eta_k(\xi_0 - 1) - \xi_k \eta_0, \tag{1.86}$$

$k = 1, 2, \ldots, n$. The geodesic flow on $T(S^n)$ is given by the Hamiltonian system

$$\boldsymbol{\xi}' = \Phi_{\boldsymbol{\eta}}, \quad \boldsymbol{\eta}' = -\Phi_{\boldsymbol{\xi}}, \tag{1.87}$$

where $' \equiv \frac{d}{ds}$,

$$\Phi = \frac{1}{2} |\boldsymbol{\xi}|^2 |\boldsymbol{\eta}|^2,$$

and Φ has the value of $\frac{1}{2}$. The time variable s is related to t by

$$t = \int |\mathbf{q}| ds, \tag{1.88}$$

which we also used in Section 1.4 for the Levi-Civita map, equation (1.36). System (1.87) yields the harmonic oscillator

$$\boldsymbol{\xi}'' + \boldsymbol{\xi} = \mathbf{0}. \tag{1.89}$$

The solutions to this system, where $\eta = \xi'$, yield the geodesic curves $\xi(s)$ on S^n. It was proven in [174] that (1.83), (1.86), (1.88) on $T_1(S_0^n)$ transform the Kepler problem (1.80) into (1.87). The energy surface $H^{-1}(-\frac{1}{2})$ is then topologically equivalent to $T_1(S_0^n)$. The north pole $\{\xi_0 = 1\}$ corresponds to collision, as one sees from (1.83) since $\{\xi_0 = 1\} \to |\mathbf{p}| = \infty$. The geodesic on S_0^n passing through the north pole therefore corresponds to a collision orbit. Restoring the north pole to the punctured sphere S_0^n, where the transformed collision orbit is smooth, therefore regularizes the flow. It is noted that a geodesic passing through the north pole $\boldsymbol{\xi}^+$ must also pass through the south pole $\boldsymbol{\xi}^-$. The south pole projects onto the origin $\mathbf{p} = \mathbf{0}$.

The above equations are for the case $h = -\frac{1}{2}$, i.e., $H = -\frac{1}{2}$. The case of arbitrary negative energy $H = h = -\frac{a}{2}, a > 0$, is $\mathbf{q} \to a\mathbf{q}, \mathbf{p} \to a^{-\frac{1}{2}}\mathbf{p}, t \to a^{\frac{3}{2}}t$.

Theorem 1.30 (Moser) *The energy surface $H^{-1}(h), h < 0$, is topologically equivalent to $T_1(S_0^n)$, where the north pole corresponds to collision states. The geodesic flow on S_0^n is mapped into the Kepler flow after a change of the independent variable.*

The idea of the proof of Theorem 1.30 outlined above relies on Hamiltonian formalism.

1.6.2 Case $h = +\frac{1}{2}$

Following the general approach of Moser in the case of $h = -1/2$, the case $h = 1/2$ was first solved in [22] along with the case of $h = 0$ and for the central repelling field. A similar construction for these cases can be found in [183]. A survey article by Milnor [163] gives a geometric description of the solutions of the various cases.

The sphere has Gaussian curvature $K(S^n) = 1$ [180]. The Gaussian curvature is a real-valued function which measures the curvature of a manifold. A formula for this is given below.

Thus, it would seem reasonable that for positive energy a topologically equivalent surface for the geodesic flow for Kepler's problem would have a constant negative curvature, $K = -1$. This is indeed the case.

We summarize the results in [22] for the cases $h = 1/2, 0$ and for the central repelling field.

Instead of S^n, we have an n-dimensional hyperboloid of two sheets $H^n_{+,-}$ ($+, -$ refer to the upper and lower sheets, respectively).

$$H^n_{+,-} = \{\boldsymbol{\xi} = (\xi_0, \xi_1, \ldots, \xi_n) \in \mathbf{R^{n+1}} | \langle \boldsymbol{\xi}, \boldsymbol{\xi} \rangle = \sum_{k=1}^{n} \xi_k^2 - \xi_0^2 = -1\},$$

embedded in a Lorentz space $\mathbf{L^{n+1}}$ defined by the metric

$$ds^2 = \sum_{k=1}^{n} d\xi_k^2 - d\xi_0^2, \tag{1.90}$$

$\langle \boldsymbol{\xi}, \boldsymbol{\eta} \rangle \equiv \sum_{k=1}^{n} \xi_k \eta_k - \xi_0 \eta_0$. It is verified that $K(H^n_{+,-}) = -1$ and that ds^2 is Riemannian when restricted to $H^n_{+,-}$; i.e., it is positive definite.

Proceeding as in the negative energy case, we find that the geodesic flow is defined by the Hamiltonian system,

$$\boldsymbol{\xi}' = \Lambda \Phi_{\boldsymbol{\eta}}, \quad \boldsymbol{\eta}' = -\Lambda \Phi_{\boldsymbol{\xi}}, \tag{1.91}$$

where $\boldsymbol{\eta} = (\eta_0, \eta_1, \ldots, \eta_n)$,

$$\Lambda = \begin{pmatrix} 1 & & \mathbf{0} \\ & \ddots & \\ \mathbf{0} & & -1 \end{pmatrix},$$

which is the identity matrix except the lower right element is -1. It is verified that (1.91) is just

$$\boldsymbol{\xi}'' - \boldsymbol{\xi} = \mathbf{0}, \tag{1.92}$$

yielding hyperbolic solutions. These are the geodesics on $H^n_{+,-} \subset \mathbf{L^{n+1}}$ and are *great hyperbolas*, obtained by intersecting $H^n_{+,-}$ with any plane passing through the origin of $\boldsymbol{\xi}$-space.

We restrict ourselves to H^n_+ and restrict $(\boldsymbol{\xi}, \boldsymbol{\eta})$ to the unit tangent bundle of H^n_+,

$$\begin{aligned} T_1(H^n_+) &= \{(\boldsymbol{\xi}, \boldsymbol{\eta}) \in \mathbf{R^{2n+2}} | \langle \boldsymbol{\xi}, \boldsymbol{\xi} \rangle \\ &= -1, \langle \boldsymbol{\eta}, \boldsymbol{\eta} \rangle = 1, \langle \boldsymbol{\xi}, \boldsymbol{\eta} \rangle = 0\}. \end{aligned}$$

The lower sheet H^n_- could have been used as well, but we use just one sheet to achieve a one-to-one correspondence with the Kepler problem. The map (1.83) is used to map $\boldsymbol{\xi} \in H^n_+$ to $\mathbf{p} \in \mathbf{R^n}$, where $\boldsymbol{\xi} \neq \boldsymbol{\xi}^+ = (1, 0, \ldots, 0)$, corresponding to the minimum point of H^n_+. Thus, $\boldsymbol{\xi} \in H^n_+ \backslash \boldsymbol{\xi}^+ \equiv H^n_{+,0}$. The map $\boldsymbol{\xi} \to \mathbf{p}$ is geometrically shown in Figure 1.11.

$\boldsymbol{\xi}^+$ is analogous to the north pole of S^n. It is seen that the great hyperbolas are mapped one-to-one onto a circle in $\mathbf{p}$-space, where part of the circle

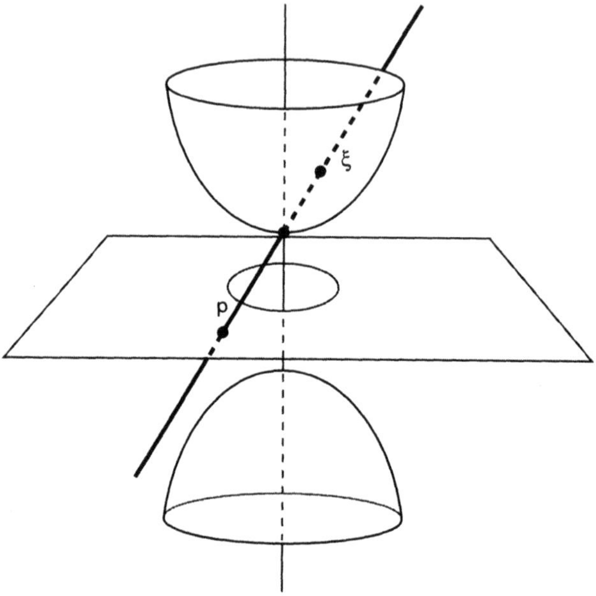

Figure 1.11 Projection from two-sheeted hyperboloid.

is missing. (See Figure 1.12.) This circle turns out to correspond to the velocity curve traced out for the Kepler problem for a hyperbolic trajectory in **q**-space. This is analogous to the case of elliptic motion. Hamilton [95] and Möbius [164] classified the velocity hodographs for the Kepler problem. They defined the term *velocity hodograph* to represent the velocity curve $\mathbf{p}(t)$ corresponding to a Kepler orbit in $\mathbf{q}(t)$.

Theorem 1.31 (Möbius, Hamilton) *The velocity hodographs for Kepler's problem are circles or parts of circles.*

It is seen that (1.83) maps $H^n_{+,0}$ isometrically into the space of $\hat{D}^n, |\mathbf{p}| > 1$, where isometric means that the induced metric ds^2 on $H^n_+ \subset \mathbf{L^{n+1}}$ is preserved in $\hat{D}^n$. $\hat{D}^n$ is a metric space which has the velocity hodographs as geodesics, which intersect $\partial\hat{D}^n, |\mathbf{p}| = 1$, normally. See Figure 1.12.

The metric on $\hat{D}^n$ is

$$ds^2 = 4(|\mathbf{p}|^2 - 1)^{-2}|\mathbf{dp}|^2, \tag{1.93}$$

which implies $K(\hat{D}^n) = 1$. This must be the case since (1.83) is isometric, and $K(H^n_+) = 1$.

A canonical extension of (1.83) is given by

$$q_k = \eta_k(\xi_0 - 1) - \xi_k\eta_0, \tag{1.94}$$

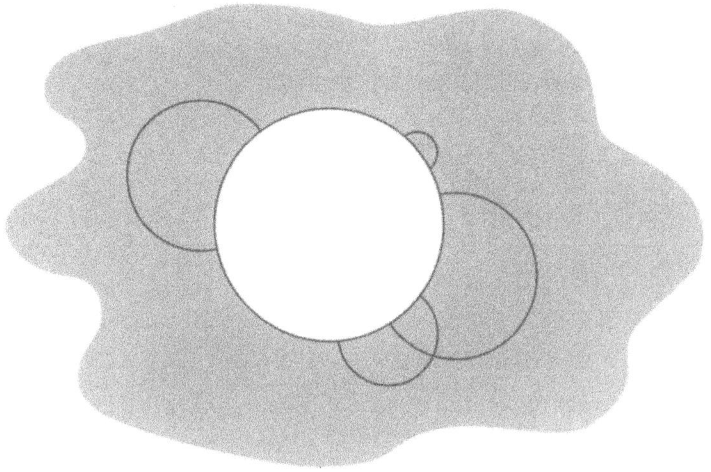

Figure 1.12 Inverted Lobachevsky disc and geodesics.

$k = 1, 2, \ldots, n$, which satisfies

$$\langle \boldsymbol{\eta}, \mathbf{d}\boldsymbol{\xi} \rangle = \langle \mathbf{y}, \mathbf{dx} \rangle, \tag{1.95}$$

$\mathbf{x} \in \mathbf{R^n}, \mathbf{y} \in \mathbf{R^n}$, and where $\mathbf{y} = \mathbf{q}, \mathbf{x} = -\mathbf{p}$. Equations (1.83), (1.94) together with (1.88) maps (1.91) into the Kepler problem (1.80) and the energy surface $H^{-1}(\frac{1}{2})$ is topologically equivalent to $T_1(H^n_{+,0})$. Thus, as stated in [22], we have the following theorem.

Theorem 1.32 *The energy surface $H^{-1}(h), h > 0$, is topologically equivalent to $T_1(H^n_{+,0}), H^n_+ \subset \mathbf{L^{n+1}}$, where the point ξ^+ corresponds to collision. The geodesic flow on $H^n_{+,0}$ is mapped one-to-one into the Kepler flow after a change of the independent variable.*

Regularization of collision is achieved in (1.91) by restoring ξ^+ to $H^n_{+,0}$ where (1.91) is smooth. Great hyperbolas passing through ξ^+ correspond to collision orbits. The case of general energy $H = h = \frac{a}{2} > 0$ is accomplished by the scaling $\mathbf{q} \to a\mathbf{q}, \mathbf{p} \to a^{-\frac{1}{2}}\mathbf{p}, t \to a^{\frac{3}{2}}t$.

It is interesting to note that inversion with respect to $\hat{D}^n, \tilde{p}_k = p_k/|\mathbf{p}|^2, k = 1, 2, \ldots, n, |\tilde{\mathbf{p}}| = |\mathbf{p}|^{-1}$, is an isometry, and when applied to (1.83) yields the map

$$\tilde{p}_k = \frac{\xi_k}{1 + \xi_0}, \qquad k = 1, 2, \ldots, n. \tag{1.96}$$

$\hat{D}^n$ is mapped into $D^n = \{|\tilde{\mathbf{p}}| < 1\}$ by the inversion, and (1.96) projects H^n_+ into D^n, as is geometrically illustrated in [22]. D^n is the classical n-dimensional Lobachevsky space, and D^2 is the well-known Lobachevsky disc.

The metric (1.93) of $\hat{D}^n$ is mapped into $d\tilde{s}^2 = 4(|\tilde{\mathbf{p}}|^2 - 1)^{-2}|\mathbf{d}\tilde{\mathbf{p}}|^2$, which agrees with (1.93), and $K(D^n) = 1$. The geodesics of D^n are the missing arcs of the geodesics of $\hat{D}^n$. See Figure 1.13.

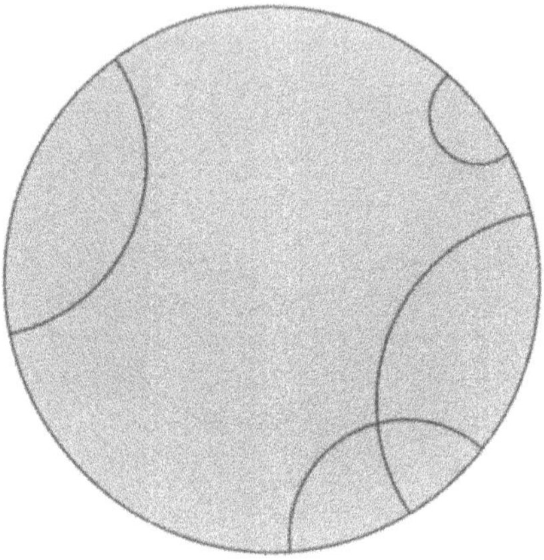

Figure 1.13 Lobachevsky disc and geodesics.

It is very interesting that the geodesics of D^n correspond to the hodograph curves *not* of the Kepler problem with the attractive gravitational force field, but of the central repelling inverse square force field, defining the *central repelling force problem,*

$$\dot{\tilde{\mathbf{q}}} = \tilde{H}_{\tilde{\mathbf{p}}}, \quad \dot{\tilde{\mathbf{p}}} = -\tilde{H}_{\tilde{\mathbf{q}}}, \tag{1.97}$$

$\tilde{\mathbf{q}} = (\tilde{q}_1, \ldots, \tilde{q}_n)$, and

$$\tilde{H} = \frac{1}{2}|\tilde{\mathbf{p}}|^2 + \frac{1}{|\tilde{\mathbf{q}}|}.$$

Collision does not occur for (1.97) for $\tilde{H} = h < \infty$.

Analogous to Theorem 1.32, we have the following theorem and corollary.

Theorem 1.33 *For a positive constant h, the energy surface $\tilde{H}^{-1}(h)$ is topologically equivalent to $T_1(H^n_+)$ embedded in $\mathbf{L}^{\mathbf{n}+1}$, and the flow of (1.97) is mapped into the geodesic flow of H^n_+ after a change (1.88) of the independent variables.*

Corollary 1.34 *The hodographs of the central repelling force problem are the geodesics of the classical Lobachevsky space. They are mapped isometrically into the hodographs of the Kepler problem for positive energy. These hodographs represent the geodesics of the inverted Lobachevsky space obtained by inversion with respect to the origin of the classical Lobachevsky space.*

More generally, we can state the following theorem [24].

Theorem 1.35 *The Kepler problem for positive energy is mapped into the central repelling problem by inversion with respect to the classical Lobachevsky space.*

Lastly, we mention the case of $h = 0$. The geodesic flow of $\mathbf{R^n}$ is given by $\boldsymbol{\xi}' = \boldsymbol{\eta}, \quad \boldsymbol{\eta}' = \mathbf{0}, \quad ' \equiv \frac{d}{ds}, \quad \boldsymbol{\xi} = (\xi_1, \ldots, \xi_n), \boldsymbol{\eta} = (\eta_1, \ldots, \eta_n)$, where $K(\mathbf{R^n}) = 0$. Let $\mathbf{R^n_0} = \mathbf{R^n} \backslash (0, 0, \ldots, 0)$ and $T_1(\mathbf{R^n}) = \{(\boldsymbol{\xi}, \boldsymbol{\eta}) \in \mathbf{R^{2n}} | |\boldsymbol{\eta}|^2 = \sum_{k=1}^{n} \eta_k^2 = 1, < \boldsymbol{\xi}, \boldsymbol{\eta} >= 0\}$, where $<, >$ is the standard Euclidean inner product.

Theorem 1.36 *The energy surface $H^{-1}(0)$ is topologically equivalent to $T^1(\mathbf{R^n_0})$, and the geodesic flow on $\mathbf{R^n}$ is mapped into the Kepler flow after a transformation of time.*

In this case the mapping into $\mathbf{p}$-space is given by

$$p_k = \frac{\xi_k}{|\boldsymbol{\xi}|^2}, \qquad k = 1, 2, \ldots, n \tag{1.98}$$

which is just inversion with respect to $\boldsymbol{\xi} = \mathbf{0}$. This maps the geodesics of $\mathbf{R^n}$, given by $\boldsymbol{\xi}'' = \mathbf{0}$, into the families of circles passing through the origin $\mathbf{p} = \mathbf{0}$, which are the hodographs. As in the other cases, (1.98) is canonically extended and time is transformed by (1.88) in order to map the geodesic flow into the Kepler flow, which is regular at collision in the geodesic coordinates, $\boldsymbol{\xi}, \boldsymbol{\eta}, s$.

1.6.3 A Simplified General Proof

A relatively short proof is given to prove the equivalence of the Kepler flow and the flow of the central repelling problem with the geodesic flows on the spaces of constant curvature for all energy cases at once.

Let $Q(a)$ be a family of n-dimensional quadratic manifolds defined by

$$Q(a) = \left\{ \mathbf{x} = (x_0, x_1, \ldots, x_n) \in \mathbf{R^{n+1}} | \sum_{k=1}^{n} x_k^2 + a^{-3} x_0^2 = a^{-1} \right\}, \tag{1.99}$$

$a \in \mathbf{R^1}$. $Q(a)$ for different a yields the following surfaces:

- For $a = 0$, $Q(a) = \mathbf{R^n}$, with coordinates $\tilde{\mathbf{x}} = (x_1, \ldots, x_n)$.

- For $a > 0$, $Q(a)$ is a family of n-dimensional ellipsoids.
- For $a < 0$, $Q(a)$ is a family of hyperboloids of two sheets.
- For $a \gg 0$, the ellipsoids are very thin, centered along the x_0-axis, with large semimajor axis. As a approaches zero, they become more spherical in shape, and for $a = 1$, $Q(a)$ is the n-sphere, S^n. For $0 < a < 1$ the sphere flattens out. As $a \downarrow 0$, $|x_0| \to 0$, and $Q(a)$ becomes infinitely thin, converging to $\mathbf{R^n}$ for $a = 0$. For $a \lesssim 0$, $\mathbf{R^n}$ bifurcates to a flattened two-sheeted hyperboloid with max and min vertice points at $x_0 = \pm a$. When $a = -1$, the standard two-sheeted hyperboloid $H^n_{+,-}$, considered previously, is obtained, and for $a \ll 0$, the two sheets become very thin, lying near the x_0-axis.

Next, embed $Q(a)$ into the respective family of spaces $\mathbf{L^{n+1}}(a)$ defined by the metric

$$d\tilde{s}^2 = \sum_{k=1}^{n} dx_k^2 + a^{-3} dx_0^2, \tag{1.100}$$

which is Riemannian.

For notation, with $\mathbf{y} = (y_0, y_1, \ldots, y_n), \mathbf{z} = (z_0, z_1, \ldots, z_n)$, we set

$$\langle \mathbf{y}, \mathbf{z} \rangle = \sum_{k=1}^{n} y_k z_k + a^{-3} y_0 z_0$$

and

$$\|\mathbf{y}\|^2 = \langle \mathbf{y}, \mathbf{y} \rangle.$$

Thus, $Q(a)$ is given by $\|\mathbf{x}\|^2 = a^{-1}$.

Lemma 1.37 *The geodesics $\mathbf{x} = \mathbf{x}(s)$ on $Q(a) \subset \mathbf{L^{n+1}}(a)$ are given by the system*

$$\mathbf{x}'' + a\mathbf{x} = \mathbf{0}, \tag{1.101}$$

$' \equiv \frac{d}{ds}$ *on the unit tangent bundle of $Q(a)$,*

$$T_1(Q(a)) = \{\, \mathbf{x} \in \mathbf{R^{n+1}}, \mathbf{x}' \in \mathbf{R^{n+1}} | \|\mathbf{x}\|^2 = a^{-1}, \\ \|\mathbf{x}'\| = 1, \langle \mathbf{x}, \mathbf{x}' \rangle = 0 \}. \tag{1.102}$$

Proof. Set $g(\mathbf{x}) = \|\mathbf{x}\|^2 - a^{-1}$. Thus, $Q(a)$ is given by $g(\mathbf{x}) = 0$. We want to find minimal paths $\mathbf{x} = \mathbf{x}(s)$ on $Q(a), s \in \mathbf{R^1}$. These are given by the *variational problem*

$$\delta \int (f - \mu g) ds = 0 \tag{1.103}$$

over all paths on $Q(a)$ connecting any two given points, where $f = \|\mathbf{x}'\|^2$ [128, 180]. The solution to this problem is given by the Euler-Lagrange equations

$$\frac{d}{dt}\left(\frac{\partial f}{\partial x_k'}\right) - \frac{\partial g}{\partial x_k} = 0, \tag{1.104}$$

$k = 0, 1, \ldots, n$. This is simplified to

$$x_k'' - \mu x_k = 0, \tag{1.105}$$

as is verified. We now determine μ. The previous equation implies

$$\langle \mathbf{x}'', \mathbf{x}\rangle - \mu\|\mathbf{x}\|^2 = 0$$

or

$$\langle \mathbf{x}'', \mathbf{x}\rangle = a^{-1}\mu. \tag{1.106}$$

On the other hand,

$$\langle \mathbf{x}', \mathbf{x}\rangle = \frac{1}{2}\frac{d}{ds}\langle \mathbf{x}, \mathbf{x}\rangle = \frac{1}{2}\frac{d}{ds}a^{-1} = 0,$$

and differentiation of this yields

$$\langle \mathbf{x}'', \mathbf{x}\rangle + \|\mathbf{x}'\|^2 = 0.$$

Equation (1.106) then implies $\mu = -a\|\mathbf{x}'\|^2$.

Since $\mathbf{x}(s)$ is a geodesic we know $\|\mathbf{x}'\| =$ constant. We choose $\|\mathbf{x}'\| = 1$, which implies s is arc-length $\tilde{s}$. Thus,

$$\mu = -a.$$

□

Lemma 1.38 $K(Q(a)) = a, Q(a) \subset \mathbf{L^{n+1}}(a)$.

Proof. We prove this for the case $n = 2$, and general $n > 2$ is an exercise.

A parametric representation of $Q(a)$ is found using geodesic coordinates by using solutions of (1.101),

$$\begin{aligned}\mathbf{x}(s) &= (x_1(s), x_2(s), x_3(s)) \\ &= a^{-\frac{1}{2}}(\sin a^{\frac{1}{2}} s, \sin a^{\frac{1}{2}}\beta \cos a^{\frac{1}{2}} s, -a^{\frac{3}{2}} \cos a^{\frac{1}{2}}\beta \cos a^{\frac{1}{2}} s)\end{aligned}$$

and the induced metric $d\tilde{s}^2$ on $Q(a)$ is given by

$$d\tilde{s}^2 = \langle \mathbf{x}_s, \mathbf{x}_s\rangle ds^2 + \langle \mathbf{x}_s, \mathbf{x}_\beta\rangle ds d\beta + \langle \mathbf{x}_\beta, \mathbf{x}_\beta\rangle d\beta^2;$$

$$\begin{aligned} g_{ss} &\equiv \langle \mathbf{x}_s, \mathbf{x}_s\rangle \equiv \langle \mathbf{x}', \mathbf{x}'\rangle = 1, \\ g_{s\beta} &\equiv \langle \mathbf{x}_s, \mathbf{x}_\beta\rangle = 0, \\ \langle \mathbf{x}_\beta, \mathbf{x}_\beta\rangle &= \cos^2 a^{\frac{1}{2}} s. \end{aligned}$$

Thus,

$$g = \begin{vmatrix} g_{ss} & g_{s\beta} \\ g_{\beta s} & g_{\beta\beta} \end{vmatrix} = \cos^2 a^{\frac{1}{2}} s.$$

The Gaussian curvature is given by [128]

$$K = -\frac{1}{2\sqrt{g}}\frac{\partial}{\partial s}\left(\frac{\frac{\partial}{\partial s} g_{\beta\beta}}{\sqrt{g}}\right) = a. \qquad \square$$

For notation, we set $\mathbf{x}^+ = (a, 0, \ldots, 0)$ and

$$Q_0(a) = Q(a)\backslash\{\mathbf{x} = \mathbf{x}^+\}.$$

If $a > 0$, $\mathbf{x}^+$ corresponds to the *north pole* of the corresponding ellipsoid. For $a < 0$, $\mathbf{x}^+$ corresponds to the minimum point of the upper sheet of the two-sheeted hyperboloid.

We consider the Kepler problem (1.80), on the energy surface $H^{-1}(h)$ given by (1.82) with $h = -a/2$.

Theorem 1.39 *The Kepler flow defined by (1.80) on the surface $H^{-1}(-\frac{a}{2})$ is mapped into the geodesic flow of (1.101) on $T_1(Q_0(a))$, where $K(Q_0(a)) = a, Q_0(a) \subset \mathbf{L^{n+1}}(a)$, and $\mathbf{x} = \mathbf{x}^+$ corresponds to collision $\mathbf{q} = 0$, which is regularized by restoring $\mathbf{x}^+$ to $Q_0(a)$. The mapping is given by*

$$p_k = \frac{a^2 x_k}{x_0 - a}, \qquad k = 1, 2, \ldots, n, \tag{1.107}$$

together with the time transformation (1.88).

Proof. Equation (1.88) applied to (1.80) implies

$$\mathbf{p}' = \frac{-\mathbf{q}}{|\mathbf{q}|^2}, \quad \mathbf{q}' = |\mathbf{q}|\mathbf{p}, \quad {}' \equiv \frac{d}{ds}. \tag{1.108}$$

Differentiation of (1.108) yields

$$\mathbf{p}'' = -|\mathbf{q}|^{-1}(\mathbf{p} + 2(\mathbf{p}, \mathbf{q})\mathbf{p}'). \tag{1.109}$$

Some basic identities are needed. Equation (1.107) implies

$$|\mathbf{p}|^2 = \frac{1}{|\tilde{\mathbf{x}}|^2}(x_0 a^{-1} + 1)^2, \tag{1.110}$$

$\tilde{\mathbf{x}} = (x_1, \ldots, x_n)$, and $\mathbf{x} = (x_0, x_1, \ldots, x_n)$. Now, $Q(a)$ can be written as $|\tilde{\mathbf{x}}|^2 = a^{-1}(1 - a^{-1}x_0)(1 + a^{-1}x_0)$, thus (1.110) becomes

$$|\mathbf{p}|^2 = a\frac{a + x_0}{a - x_0}. \tag{1.111}$$

It is noted that (1.111) yields the *inverse map*, using (1.107), which yields

$$x_0 = \frac{|\mathbf{p}|^2 - a}{1 + a^{-1}|\mathbf{p}|^2}, \qquad x_k = -2\frac{a^{-1}p_k}{1 + a^{-1}|\mathbf{p}|^2}, \tag{1.112}$$

$k = 1, 2, \ldots, n$. (1.111) is a map of the norm of $\mathbf{p}$. To obtain a map of $|\mathbf{q}|$, (1.82) implies $|\mathbf{q}|^{-1} = \frac{1}{2}(a+|\mathbf{p}|^2)$, which upon substitution into (1.111) yields

$$|\mathbf{q}| = a^{-2}(a - x_0). \tag{1.113}$$

Finally, differentiation of (1.113) and using (1.108) yields

$$(\mathbf{p}, \mathbf{q}) = -a^{-2}x_0', \tag{1.114}$$

$(\mathbf{p}, \mathbf{q}) = p_1q_1 + \cdots + p_nq_n$.

Equation (1.107) together with the identities (1.113), (1.114) is all we need to transform (1.109).

Differentiating (1.107) twice with respect to s yields

$$p_k' = A_k(x_0 - a)^{-2}, \tag{1.115}$$

$$p_k'' = \frac{(x_0 - 1)(x_k''a^2(x_0 - a) - a^2x_kx_0'') - 2A_kx_0'}{(x_0 - a)^3}, \tag{1.116}$$

where $A_k = (x_0 - a)a^2x_k' - a^2x_kx_0'$. Substituting (1.113), (1.114), (1.107), (1.115) into the right-hand side of (1.109) yields

$$p_k'' = \frac{a^4x_k(x_0 - a) - 2x_0'A}{(x_0 - a)^3}. \tag{1.117}$$

Finally, equating (1.116), (1.117) yields

$$ax_k'' + a^2x_k = x_k''x_0 - x_kx_0'', \tag{1.118}$$

$k = 1, 2, \ldots, n$. This equation is used to conclude the proof by taking inner products. We first show $x_0'' + ax_0 = 0$. Take the inner product of (1.118) with $\tilde{\mathbf{x}}$,

$$a(\tilde{\mathbf{x}}'', \tilde{\mathbf{x}}) + a^2|\tilde{\mathbf{x}}|^2 = (\tilde{\mathbf{x}}'', \tilde{\mathbf{x}})x_0 - |\tilde{\mathbf{x}}|^2x_0''. \tag{1.119}$$

Since $\mathbf{x}, \mathbf{x}' \in T_1(Q_0(a)), |\tilde{\mathbf{x}}|^2 = a^{-1} - a^{-3}x_0^2$ we get an expression for $(\tilde{\mathbf{x}}'', \tilde{\mathbf{x}})$ by differentiating $\langle \mathbf{x}, \mathbf{x}' \rangle = 0$ with respect to s and using $\|\mathbf{x}'\|^2 = 1$, which together yield $(\tilde{\mathbf{x}}, \tilde{\mathbf{x}}'') = -1 - a^{-3}x_0x_0''$. Equation (1.119) becomes

$$(x_0'' + ax_0)(a^{-1}x_0 - 1) = 0.$$

But $\mathbf{x} \neq \mathbf{x}^+$, so that $a^{-1}x_0 - 1 \neq 0$, and we obtain

$$x_0'' + ax_0 = 0.$$

To obtain $\tilde{\mathbf{x}}'' + a\tilde{\mathbf{x}} = \mathbf{0}$, we substitute $x_0'' = -ax_0$ into (1.118). This implies $(\tilde{\mathbf{x}}'' + a\tilde{\mathbf{x}})(a - x_0) = 0$, yielding

$$\tilde{\mathbf{x}}'' + a\tilde{\mathbf{x}} = \mathbf{0}. \qquad \square$$

The case of the central repelling problem is obtained by considering more generally the Hamiltonian system (1.80) with Hamiltonian

$$H = \frac{1}{2}|\mathbf{p}|^2 - \frac{\Delta}{|\mathbf{q}|} \tag{1.120}$$

on the surface $H^{-1}(-\frac{a}{2}), \Delta \in \mathbf{R^1}$. This case is obtained as a scaling

$$\mathbf{p} \to \Delta^{-\frac{1}{3}}\mathbf{p}, \quad \mathbf{q} \to \Delta^{-\frac{1}{3}}\mathbf{q}, \quad \mathbf{a} \to \Delta^{-\frac{2}{3}}\mathbf{a},$$

in all the previous equations in Theorem 1.39 and its proof. Equation (1.101) is replaced by

$$\mathbf{x}'' + a\Delta^{-\frac{2}{3}}\mathbf{x} = 0,$$

$(\mathbf{x}, \mathbf{x}') \in T_1(Q(a\Delta^{-\frac{2}{3}}))$. For example,

$$H = \frac{1}{2}|\mathbf{p}|^2 - \frac{1}{|\mathbf{q}|} = -\frac{a}{2}$$

becomes

$$\tilde{H} = \frac{1}{2}\Delta^{-\frac{2}{3}}|\mathbf{p}| - \frac{\Delta^{\frac{1}{3}}}{|\mathbf{q}|} = -\frac{a}{2}\Delta^{-\frac{2}{3}}$$

and multiplying by $\Delta^{\frac{2}{3}}$ yields (1.120) with $H \equiv \tilde{H}\Delta^{\frac{2}{3}}$. Equation (1.80) is seen to become $\dot{\mathbf{q}} = \mathbf{p}, \dot{\mathbf{p}} = -\Delta\mathbf{q}|\mathbf{q}|^{-3}$.

The results obtained in this section are for the two-body problem. More generally, we could ask if geodesic equivalent flows can be constructed for the $n-$body problem for $n \geq 3$. This was recently answered in an interesting paper by McCord, Meyer, and Offin [149], where, in general the answer is no unless the angular momentum is zero and the energy is positive.

Chapter Two

Bounded Motion, Cantor Sets, and Twist Maps

Quasi-periodic motion is considered in section 2.1, and it is described relative to the planar circular restricted three-body problem. The KAM theorem is applied to this problem to prove the existence of quasi-periodic motion on invariant tori for the motion of the zero mass particle about the larger primary. In section 2.2 the flow on the tori is reduced to the consideration of a two-dimensional area-preserving map and on the existence of invariant curves for this map. The map is near to a rotation and falls under the category of what is called a monotone twist map. The Moser twist theorem is used to prove the existence of invariant curves in section 2.2. Cantor sets are discussed in subsection 2.2.1. These sets are always associated with the chaotic dynamics considered in this book and therefore play a key role.

In section 2.3 area-preserving maps with fixed points are considered. Normal forms for these maps are described in both the elliptic and hyperbolic cases. In the hyperbolic case, the stable manifold theorem is discussed with relevance to the existence of local stable and unstable manifolds. The global dynamics associated to these manifolds is discussed, and homoclinic and heteroclinic points are defined. The complicated dynamics that results due to the existence of transverse homoclinic and heteroclinic points is described. It is proven that the normal form for the map with an elliptic fixed point is equivalent to a monotone twist map. The general nature of the complicated dynamics in a neighborhood of an elliptic fixed point is described and related to a theorem of Zehnder.

In section 2.4 a proof is given of the existence of the *periodic orbits of the first kind* first proven to exist by Poincaré. This is carried out in detail. The plan of this section follows that of the book by Siegel and Moser [204], and we expand on that presentation with some added calculations and results including the explicit solutions of the variational equations relative to circular orbits for the two-body problem. Several different methods to do this are described. Once the family of periodic orbits of the first kind are proven to exist, they can be used to construct an area-preserving map in the neighborhood of an elliptic fixed point. This map plays a key role in the remainder of this chapter. It's first order approximation is explicitly determined. With this map the Moser twist theorem can be applied to prove the existence of invariant curves about the fixed point.

Following [204] we apply the Birkhoff fixed point theorem to prove the existence of periodic points in the resonance gaps where the Moser twist theorem is not applicable. The resulting dynamics near the periodic orbits of the first kind due to the existence of these points is described. This is done in section 2.5.

In section 2.5 much more is said about the dynamics of M in the resonance gaps. This is accomplished by the application of Aubrey-Mather theory, which is described in this section. Aubrey-Mather sets are defined. We prove the result that generally there exists Aubrey-Mather invariant sets in any sufficiently small neighborhood of a given periodic orbit of the first kind of the restricted problem on a two-dimensional surface of section. This is stated in Theorem 2.33.

2.1 QUASI-PERIODICITY AND THE KAM THEOREM

We consider the planar restricted problem (1.61) in rotating coordinates.

For $\mu \ll 1$, a great deal is known about the nature of the dynamics of bounded motion of P_3 about P_1 with initial negative Kepler energy. Assume that $\mu = 0$, that P_3 moves about P_1 in a Kepler ellipse of angular frequency $\tilde{\omega}, \tilde{\omega} = 2\pi/T$, where T is the period of motion of P_3 about P_1. Let ω be the frequency of P_2 about P_1 we have normalized to 1. (Under our normalizations this implies the angular velocity of P_2 is 1.) In the rotating coordinate system, the elliptical orbit of P_3 precesses with frequency $\omega = 1$. (See Figure 2.1.)

Consider the ratio $\tau = \tilde{\omega}/\omega$. We will write ω in the formulas for generality, however it will have the normalized value of 1. If $\tilde{\omega} \in \mathbb{Z}^+, \mathbb{Z}^+$ is the set of positive integers, then P_3 is in a periodic orbit about P_1. For example, if $\tilde{\omega} = 2$, then P_3 will go around P_1 twice in the time it takes P_2 (now of mass zero) to go around P_1 once. The motion of P_3 is in *resonance* with the motion of P_2, and the orbit of P_3 is called a *resonance orbit* of *type* $(2, 1)$. More generally, if τ is a positive rational number m/n, then P_3 is in resonance with P_2 and the resonances are defined as *type* (m, n), $m \geq 1, n \geq 1, m, n \in \mathbb{Z}^+$. When τ is irrational, then P_3 is not periodic. Its motion is called *quasi-periodic*, with two frequencies $\omega, \tilde{\omega}$.

Now assume that the Jacobi constant $\tilde{J} = C$ is fixed, and set $\mu = 0$. The motion of P_3 lies on the three-dimensional surface $\tilde{J}^{-1}(C)$ given by

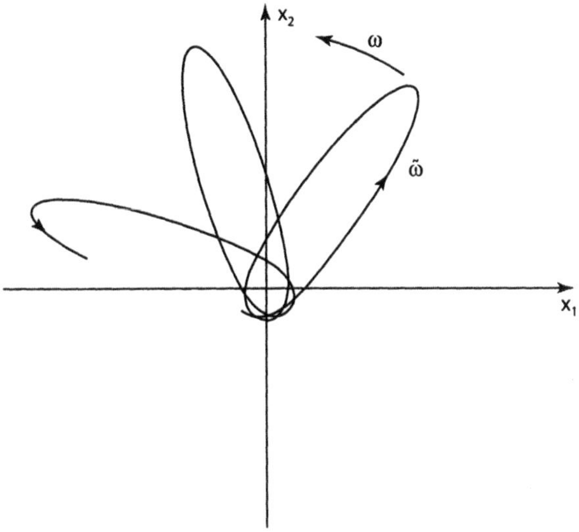

Figure 2.1 Precessing elliptical motion.

(1.64) with $\mu = 0$. As τ varies on the set of irrational numbers, a family of quasi-periodic orbits, parameterized by the two-body energy $\tilde{H} < 0$, and angular momentum in rotating coordinates $\tilde{L}$ are obtained, each of which are constant for $\mu = 0$, where by (1.71) $\tilde{J} = -2[\tilde{H} - \tilde{L}]$. $\tilde{H}$ yields the frequency $\tilde{\omega}$ through Kepler's third law, $T^2 = (2\pi)^2 a^3$. $\tilde{L}$ is associated to the uniform rotation of the coordinate system, $\tilde{H} = -\frac{1}{2a}, \tilde{L} = \sqrt{a(1-e^2)}$.

As τ, or equivalently $\tilde{\omega}$, varies, ellipses of different semimajor axis are obtained of two frequencies $\tilde{\omega}, \omega = 1$. For each value of $\tilde{J} = -2(\tilde{H} - \tilde{L}) = C$, as $\tilde{H}, \tilde{L}$ vary, or equivalently as a, e vary, the appearance of the family of the precessing ellipses changes. This family depends on $\tilde{H}$ and $\tilde{L}$, or equivalently on the values of a, e. These can be used to therefore identify a particular type of precessing elliptic motion on $\tilde{J}^{-1}(C)$, which precesses with uniform frequency ω.

Once a, e are chosen, the frequency $\tilde{\omega}$ of motion of P_3 about P_1 is therefore determined as $\tilde{\omega}(a)$. Depending on the value of a, the motion of P_3 is either quasi-periodic of frequencies $\tilde{\omega}, \omega$, or resonant. These two frequencies can be used as coordinates on a two-dimensional torus $T^2(a, e)$ or equivalently $T^2(\tilde{H}, \tilde{L})$. For irrational values of $\tilde{\omega}$, the motion of P_3 is *dense* on the corresponding torus. This means that if one chooses any open neighborhood of any point on T^2, no matter how small, if the time of motion is sufficiently large, the trajectory of P_3 on T^2 will pass though this neighborhood. The angle $2\pi\tilde{\omega}t$ measures the latitude of T^2, and $2\pi\omega t$ measures the longitude. As t varies, P_3 winds around T^2.

We can use a, e to get an intuitive feel for the dimensions on the tori. First, they are in phase space on $\tilde{J}^{-1}(C)$ and surround P_1. When a is very small, the periapsis distance $r_p = a(1-e)$ is close to P_1, as is the apoapsis distance, for all $e \in [0,1)$. These are therefore small tori and the frequency $\tilde{\omega}(a)$ is therefore very large since $\tilde{\omega} = a^{-\frac{3}{2}}$. That is, the velocity of P_3 about P_1 is very large since P_3 is near collision. The magnitudes of the periapsis velocity and apoapsis velocity are given by $\sqrt{\frac{1+e}{r_{13}}}, \sqrt{\frac{1-e}{r_{13}}}$, respectively, in inertial coordinates for $r_{13} \ll 1$. Thus, P_3 performs many latitude winding cycles on T^2 in one longitude period $2\pi\omega$. As a increases, the tori become larger since the apoapsis distance $r_a = a(1+e)$ will increase. The family of tori are concentric, so that one lies within the other. The tori give rise to either resonant motion or quasi-periodic motion. We refer to them as *resonant tori and irrational tori*, respectively. The motion of P_3 is constrained to lie on the tori for all time, and the velocity vector of P_3 in phase space is tangent to the tori. Thus, the tori are called *invariant*.

For the respective motions on the tori, P_3 in physical space is moving in precessing ellipses about P_1. In inertial space, there is just a single ellipse. Thus, in inertial phase space, the tori collapse to individual periodic orbits.

When $\mu \neq 0$ and μ is sufficiently small, it is intuitively clear that the motion of P_3 on the precessing ellipse is likely not to change too much if P_3 is not significantly gravitationally perturbed by P_2. For example, if P_3 is an asteroid moving about the Sun and Jupiter is the perturbation where $\mu = 0.001$, and the asteroid's orbit does not come too close to Jupiter, then the orbit of the asteroid will approximately be a Kepler ellipse, where a and e change very little as time progresses.

A more mathematical way of stating this is that under small perturbations the invariant tori generally survive, and geometrically their dimensions change by very little. This is approximately what happens, provided the value of $\tilde{\omega}$ is not too close to being an integer. That is, the motion of P_3 is not too close to resonant motion. This is the essence of the KAM theorem in this particular situation. It has many variations and is applicable to a wide class of Hamiltonian systems. We will state it here in a version relevant to our discussion.

Let $H(\mathbf{x}, \mathbf{y}, \mu)$ be a general Hamiltonian function where

$$\dot{\mathbf{x}} = H_{\mathbf{y}}, \qquad \dot{\mathbf{y}} = -H_{\mathbf{x}} \tag{2.1}$$

are the corresponding differential equations, $\mathbf{x} \in \mathbf{R^2}, \mathbf{y} \in \mathbf{R^2}$, and μ is in a small neighborhood of 0. H is assumed to be smooth in the variables $x_k, y_k, \mu, k = 1, 2,$ and periodic of period 2π in x_k. $\mathbf{y} \in \mathcal{D}, \mathcal{D} \subset \mathbf{R^2}$ is an open set, $\mathbf{x} = (x_1, x_2), \mathbf{y} = (y_1, y_2)$. Assume H is of the form

$$H = H_0(\mathbf{y}) + H_1(\mathbf{x}, \mathbf{y}, \mu), \tag{2.2}$$

$H_1(\mathbf{x}, \mathbf{y}, 0) = 0$. Let $H^{-1}(h)$ represent the set

$$H^{-1}(h) = \{\mathbf{x}, \mathbf{y} | H = h, h \in \mathbf{R^1}\}.$$

We restrict the flow of (2.1) to the three-dimensional set $H^{-1}(h)$. For $\mu = 0$, (2.1) implies

$$\dot{\mathbf{x}} = H_{0_\mathbf{y}}(\mathbf{y}), \quad \dot{\mathbf{y}} = \mathbf{0}, \tag{2.3}$$

which gives

$$\mathbf{y} = \mathbf{c}_0 = \text{constant}, \quad \mathbf{x}(t) = \boldsymbol{\omega} t + \mathbf{x}(0), \tag{2.4}$$

$\boldsymbol{\omega} = H_{0\mathbf{y}}(\mathbf{c}_0) =$ constant vector. ω_k represent the frequencies of motion $\boldsymbol{\omega} = (\omega_1, \omega_2)$. This motion is defined by $\mathbf{x}(t), t \in \mathbf{R^1}$, and can be viewed as lying on an invariant torus defined by identifying x_1, x_2 modulo 2π, and parameterized by $\mathbf{c}_0$. We label the family of tori $T^2(\mathbf{c}_0)$. The flow on $T^2(\mathbf{c}_0)$ is given by lines with slope ω_2/ω_1. (See Figure 2.2.)

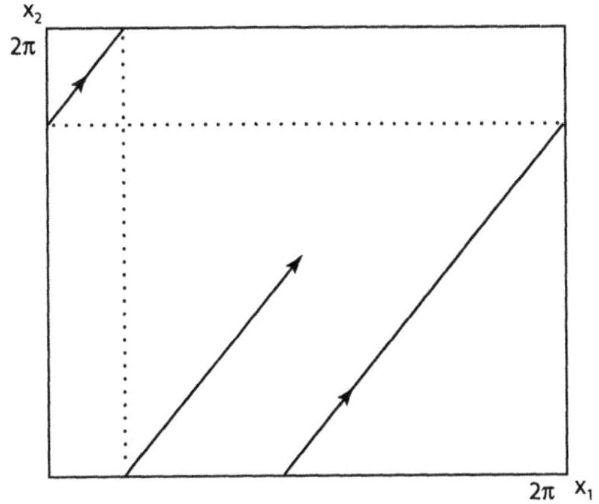

Figure 2.2 Flow on torus.

As $\mathbf{c}_0$ varies in $\mathcal{D}$, $\boldsymbol{\omega}$ changes. When $\omega_2/\omega_1 = m/n, n \neq 0$, then a single periodic orbit exists on T^2. When ω_2/ω_1 is irrational, then the curve $\mathbf{x}(t)$ is dense on T^2 [15, 16]. In this case $\mathbf{x}(t)$ is quasi-periodic of the two frequencies ω_1, ω_2. ω_1, ω_2 are analogous to $\omega, \tilde{\omega}$, respectively.

Theorem 2.1 (Kolmogorov, Arnold, Moser) *(See [204].) For μ sufficiently small, assume*

$$\begin{vmatrix} H_{0y_1y_1} & H_{0y_1y_2} & H_{0y_1} \\ H_{0y_2y_1} & H_{0y_2y_2} & H_{0y_2} \\ H_{0y_1} & H_{0y_2} & 0 \end{vmatrix} \neq 0, \tag{2.5}$$

where $y \in \mathcal{D}$; then on each surface $H^{-1}(h)$ there exists a set of positive measure of quasi-periodic solutions of the form

$$\mathbf{x}(t) = \boldsymbol{\omega} t + \mathbf{x}(0) + \mathbf{F}(\mathbf{x}(t), \mathbf{x}(t), \mu),$$
$$\mathbf{y}(t) = \mathbf{c}_0 + \mathbf{G}(\mathbf{x}(t), \mathbf{y}(t), \mu); \tag{2.6}$$

$\mathbf{F}(\mathbf{x}(t), \mathbf{y}(t), 0) = \mathbf{G}(\mathbf{x}(t), \mathbf{y}(t), 0) = \mathbf{0}$. $\mathbf{F}, \mathbf{G}$ have period 2π in $\mathbf{x}, \boldsymbol{\omega} = H_{0_\mathbf{y}}(\mathbf{c}_0)$ and are smooth functions of $\mathbf{x}, \mathbf{y}, \mu$. These solutions lie on invariant tori T^2 parameterized by $\mathbf{c}_0$, which specify the tori.

Thus, (2.6) is a smooth continuation of (2.4), and therefore, there exists a set of positive measure of invariant tori $T^2(\mathbf{c}_0, \mu)$ as perturbations from the case of $\mu = 0$, for each fixed $\mu > 0$, as $\mathbf{c}_0$ varies. $T^2(\mathbf{c}_0, \mu)$ are called *KAM tori.*

Condition (2.5) guarantees that on $H^{-1}(h)$ the frequencies ω_1, ω_2 can be chosen so that they are sufficiently bounded far enough from rational numbers. More precisely, they satisfy

$$|j_1\omega_1 + j_2\omega_2| \geq \hat{c}|j|^{-\delta} \tag{2.7}$$

for all integers j_k and where $|j| = |j_1| + |j_2| > 0$, where $\hat{c}, \delta$ are positive constants. Condition (2.5) is called *iso-energetic nondegeneracy* since we are restricting the flow to a fixed energy surface.

The set of positive measure on $H^{-1}(h)$ can be made more precise. Let $(\mathbf{x}, \mathbf{y}), \mathbf{y} \in \mathcal{D}$, be a three-dimensional open set S of points on $H^{-1}(h)$, and let ϵ be a given positive constant. Then, there exists a value $\mu_0 = \mu_0(\epsilon)$ such that for $|\mu| < \mu_0$ the quasi-periodic solutions (2.6) fill out a closed set $\tilde{S}$, where the measure m of $\tilde{S}$ is given by

$$m(S - \tilde{S}) < \epsilon m(S). \tag{2.8}$$

Thus, since ϵ can be made arbitrarily small most of the solutions of S will be quasi-periodic, lying on invariant tori.

Returning to the motion of P_3, to prove that its motion is quasi-periodic of frequencies $\tilde{\omega}, \omega$, which are analogous to ω_2, ω_1, respectively, in Theorem 2.1, we need to represent the system (1.61) as a Hamiltonian system and then insure that it can be put in the special form (2.2). It is seen that this form is distinguished by the fact that the unperturbed term $H_0(\mathbf{y})$ does not depend on the variables $x_k, k = 1, 2$, and that H_1 is periodic in x_k and vanishes when $\mu = 0$.

The fact that H_1 vanishes when $\mu = 0$ means that the Hamiltonian system is close to the one given by H_0 when μ is small. On the other hand, the system with Hamiltonian H_0 can be explicitly solved; in other words, it is *integrable.* Thus (2.1) represents a *nearly integrable* Hamiltonian system. This is the type of system KAM theory generally addresses. Thus, the choice of the variables used to achieve the special form of H of (2.2) is key. For this reason they are given a special name, *action, angle variables.* The y_k are the action and x_k are the angle variables. Once action, angle variables are found, then Theorem 2.1 can be immediately applied, assuming the required conditions are satisfied.

Now, (1.61) is not in Hamiltonian form. In chapter 1, it is shown how to transform (1.61) into Hamiltonian form. This is done by (1.75),

$$z_k = x_k, \qquad w_1 = \dot{x}_1 - x_2, \qquad w_2 = \dot{x}_2 + x_1, \tag{2.9}$$

which yields the Hamiltonian system,

$$\begin{aligned} \dot{\mathbf{z}} &= \tilde{H}_{\mathbf{w}}, \qquad \dot{\mathbf{w}} = -\tilde{H}_{\mathbf{z}}, \\ \tilde{H} &= \frac{1}{2}|\mathbf{w}|^2 - z_1 w_2 + z_2 w_1 - \tilde{\Omega}(z,\mu), \end{aligned} \tag{2.10}$$

$\tilde{\Omega} = \mu w_2 + \frac{1-\mu}{|\mathbf{z}|} + \frac{\mu}{|\mathbf{z}-\mathbf{e}_1|}, \mathbf{e}_1 = (1,0)$, where we perform the translation $z_1 \to z_1 + \mu$ which places P_1 at the origin. $\tilde{H}$ corresponds to G in (1.74). The term

$$\hat{L} = z_1 w_2 - z_2 w_1$$

corresponds to the angular momentum in *rotating canonical variables.* For $\mu = 0$, (2.10) becomes

$$\tilde{H} \equiv H_0(\mathbf{z},\mathbf{w}) = \hat{H}_0(\mathbf{z},\mathbf{w}) - \hat{L}(\mathbf{z},\mathbf{w}), \tag{2.11}$$

where

$$\hat{H}_0(\mathbf{z},\mathbf{w}) = \frac{1}{2}|\mathbf{w}|^2 - \frac{1}{|\mathbf{z}|},$$

analogous to (1.71). Both $\hat{H}_0, \hat{L}$ are integrals for $\mu = 0$, corresponding to the Kepler energy and angular momentum integrals. $\hat{L}$ corresponds to the uniform rotation of the coordinate system of frequency $\omega = 1$.

For Kepler motion of P_3 about P_1 for $\mu = 0$, the value of $\hat{H}_0, \hat{L}$ in inertial coordinates is preserved by (1.29):

$$\hat{H}_0 = -\frac{1}{2a}, \qquad \hat{L} = \sqrt{a(1-e^2)}.$$

The action, angle coordinates we transform to will use the orbital elements of the elliptic motion, i.e., a, e, eccentric anomaly E, and true anomaly θ, as the new coordinates. The transformation will canonically map $z_1, z_2, w_1, w_2 \to x_1, x_2, y_1, y_2$, where the x_k, y_k are functions of a, e, E, θ. These canonical variables are called *Delaunay variables.* In order for them to be well defined it will be necessary to restrict $0 < e < 1$. These variables x_1, x_2, y_1, y_2 are not to be confused with x_k, y_k in (1.61). The first set of Delaunay variables are

$$y_1 = \sqrt{a}, \qquad y_2 = y_1\sqrt{1-e^2}, \tag{2.12}$$

where we require $e < 1$. Thus, in these coordinates,

$$\hat{H}_0 = -\frac{1}{2y_1^2}, \qquad \hat{L} = y_2, \tag{2.13}$$

where for notational simplicity we use the symbols $\hat{L}, \hat{H}_0$, which were used previously in the variables $\mathbf{z}, \mathbf{w}$, also for functions of x_k, y_k without confusion.

Thus, for the case of $\mu = 0$,

$$H_0 = H_0(\mathbf{y}) = \frac{-1}{2y_1^2} - y_2. \tag{2.14}$$

y_1, y_2 will correspond to the action variables.

The second set of Delaunay variables are

$$\begin{aligned} x_1 &= \mathcal{M} = a^{-3/2}(t - t_0) = E - e \sin E, \\ x_2 &= \omega t, \qquad \omega = 1, \end{aligned} \tag{2.15}$$

where $\mathcal{M}$ is the mean anomaly defined in section 1.3, and E is the eccentric anomaly. A canonical transformation D is constructed.

D can be explicitly constructed using a *generating function.* A generating function is one that allows the computation of half of the components of a canonical transformation from one-half of the known components. That is, knowing one-half of the desired transformation allows the computation of the other half. In the process of doing this, the required generating function can be obtained through a partial differential equation called the *Hamilton-Jacobi equation.* Often, though, the generating function can be made from observation. This is a lengthy subject area, and the reader should consult the literature [214].

From (1.46),

$$\begin{aligned} z_1 &= h_1(a, E, e) = a(\cos E - e), \\ z_2 &= h_2(a, E, e) = a\sqrt{1 - e^2} \sin E, \end{aligned} \tag{2.16}$$

$a > 0$. These are just the solutions of the Kepler problem in inertial coordinates. The most general solution to the Kepler problem in rotating coordinates is obtained by multiplying $\mathbf{z} = (z_1, z_2)^T$ by $R(t)$ given by (1.59), so that more generally,

$$\begin{aligned} z_1 &= z_1(a, e, E, t) = h_1 \cos t + h_2 \sin t, \\ z_2 &= z_2(a, e, E, t) = -h_1 \sin t + h_2 \cos t; \end{aligned} \tag{2.17}$$

$(z_1, z_2)^T$ represents the transpose. The variables e, a, E, t are expressed in terms of y_1, y_2, x_1, x_2 using (2.12), (2.15). That is,

$$a = y_1^2, \quad e = \sqrt{1 - (y_2/y_1)^2}, \quad t = x_2,$$

and from $x_1 = E - e \sin E$, we implicitly solve for $E = E(e, x_1) = E(y_1, y_2, x_1)$. Thus

$$\begin{aligned} z_1 &= g_1(\mathbf{x}, \mathbf{y}) = h_1 \cos x_2 + h_2 \sin x_2, \\ z_2 &= g_2(\mathbf{x}, \mathbf{y}) = -h_1 \sin x_2 + h_2 \cos x_2, \end{aligned} \tag{2.18}$$

where

$$\begin{aligned} h_1 &= y_1^2 \left(\cos E(y_1, y_2, x_1) - \sqrt{1 - (y_2/y_1)^2} \right), \\ h_2 &= y_1 y_2 \sin E(x_1, y_2, x_2), \end{aligned} \tag{2.19}$$

represent one-half of the transformation of Delaunay variables, $z_1, z_2 \to x_1, x_2, y_1, y_2$. The canonical extension to w_1, w_2 is obtained in [214] and is given by

$$w_k = y_1^{-3} g_{k_{x_1}}, \tag{2.20}$$

$k = 1, 2$. It is verified that

$$\sum_{k=1}^{4} dz_k dw_k = \sum_{k=1}^{4} dx_k dy_k.$$

D is given by (2.18), (2.20). From (2.13), (2.11), D maps (2.10) into

$$H(\mathbf{x}, \mathbf{y}) \equiv \tilde{H}(\mathbf{z}(\mathbf{x}, \mathbf{y}), \mathbf{w}(\mathbf{x}, \mathbf{y})) = H_0(\mathbf{y}) + H_1(\mathbf{x}, \mathbf{y}, \mu), \tag{2.21}$$

where

$$H_0(\mathbf{y}) = -\frac{1}{2y_1^2} - y_2, \tag{2.22}$$

$$H_1(\mathbf{x}, \mathbf{y}, \mu) = -\mu \left[y_1^{-3} g_{2_{x_1}} - \frac{1}{|\mathbf{h}|} + \frac{1}{|\mathbf{h} - \mathbf{e}_1|} \right]; \tag{2.23}$$

$\mathbf{h}$ is given by (2.19). The differential equations are

$$\dot{\mathbf{x}} = H_{\mathbf{y}}, \quad \dot{\mathbf{y}} = -H_{\mathbf{x}}. \tag{2.24}$$

Equation (2.21) is in the required form (2.2). We need to restrict y_k so that collision of P_3 with P_2 does not occur. Assume first that the precessing ellipses lie completely interior to the motion of P_2 about P_1. This means

$$r_a = a(1 + e) < 1. \tag{2.25}$$

It can be shown that this is equivalent to $y_1 < (2 - y_2^2)^{-\frac{1}{2}} < 1$, which we prove in Lemma 2.2. Also, $a > 0, 0 < e < 1$, or $0 < |y_2| < y_1, y_2 > 0$ represents counterclockwise motion as t increases, and $y_2 < 0$ represents clockwise motion. $e \neq 0$ since x_2 is not defined for $e = 0$. Thus, the variables $\mathbf{y}$ are restricted to the domain,

$$\mathcal{D} = \left\{ \mathbf{y} | 0 < |y_2| < y_1 < (2 - y_2^2)^{-\frac{1}{2}} \right\}. \tag{2.26}$$

$\mathcal{D}$ has two components, one for counterclockwise motion and the other for clockwise motion.

Lemma 2.2 *Equation (2.25) is equivalent to* $y_1 < (2 - y_2^2)^{-\frac{1}{2}} < \frac{1}{2}$.

Proof. Equation (2.12) implies $e = (1 - y_2^2 y_1^{-2})^{-\frac{1}{2}}$. Since $r_a > 0$, (2.25) then implies

$$0 < y_1^2(1 + [1 - y_2^2 y_1^{-2}]^{\frac{1}{2}}) < 1.$$

Dividing by $y_1^2 > 0$, adding -1, and squaring gives

$$1 < 1 - y_2^2 y_1^{-2} < (y_1^{-2} - 1)^2.$$

Adding -1 implies

$$0 < -y_2^2 < y_1^{-2} - 2.$$

Finally, adding 2 implies

$$\frac{1}{2} > (2 - y_2^2)^{-\frac{1}{2}} > y_1. \qquad \square$$

Our estimate of $\mathcal{D}$ is slightly sharper than that of [204]. We are also considering the set where

$$r_p = a(1 - e) > 1, \tag{2.27}$$

which is not considered in [204]. This also insures the motion of P_3 for $\mu = 0$ will not collide with P_2, $m_2 = 0$. Equation (2.27) can be written as

$$0 < (2 - y_2^2)^{-\frac{1}{2}} < y_1. \tag{2.28}$$

The proof is similar to that of Lemma 2.2. We therefore define another set with two components,

$$\tilde{\mathcal{D}} = \left\{ \mathbf{y} | 0 < |y_2| < (2 - y_2^2)^{-\frac{1}{2}} < y_1 \right\}. \tag{2.29}$$

Set $\hat{\mathcal{D}} = \mathcal{D} \cup \tilde{\mathcal{D}}$.

It is verified that the nondegeneracy condition (2.5) is satisfied on the set $\hat{\mathcal{D}}$, as a straightforward computation shows. In fact, this determinant D turns out to be

$$D = \begin{vmatrix} -3y_1^{-4} & 0 & y_1^{-3} \\ 0 & 0 & -1 \\ y_1^{-3} & -1 & 0 \end{vmatrix} = 3y_1^{-4} \neq 0.$$

Thus, the assumptions of Theorem 2.1 are satisfied and we conclude that there is a relatively large set of positive measure of quasi-periodic motion for P_3 on $\tilde{J}^{-1}(C)$. We summarize this in the following theorem.

Theorem 2.3 *For $\mu > 0$ sufficiently small there exists a set of positive measure of quasi-periodic orbits for P_3 in the set $\hat{\mathcal{D}}$ lying on two-dimensional invariant tori with two frequencies $\tilde{\omega} = a^{-\frac{3}{2}}, \omega = 1$. The orbits of (1.61) lie either totally within the orbit of P_2 or totally outside of it, and appear as precessing ellipses. The frequencies $\tilde{\omega}, \omega$ are nonrational and satisfy (2.7) with $\omega_1 = 1, \omega_2 = \tilde{\omega}$. The set of orbits where (2.7) is not satisfied is a set of small measure estimated by (2.8).*

The dynamics of the orbits in the excluded set $\mathcal{R} = S - \tilde{S}$ is much more complicated than the motion on the invariant tori. It is discussed in sections 2.3 and 2.5.

2.1.1 Stability of Motion

The motion of P_3 is stable when moving on the KAM tori as guaranteed by Theorem 2.3. This is because the action variables y_1, y_2 or equivalently a, e will change by $\mathcal{O}(\mu)$, when $\mu \ll 1$, for all time. This *orbital stability* of P_2 also occurs if P_3 moves in the excluded region $\mathcal{R}$. This fact has to do with dimensional considerations. $\mathcal{R}$ occurs as gaps between the concentric invariant tori where, as we will see, the motion is complicated. However, the gaps are very thin, and for P_3 moving in this region it is squeezed between neighboring invariant tori. P_3 cannot move away from these two neighboring tori as it is trapped for all time between them. This is because the tori are two-dimensional and the surface $\tilde{J}^{-1}(C)$ on which they lie is three-dimensional. Thus, one dimension remains, and this means the tori trap P_3. This is analogous to having a particle lying between two concentric circles on $\mathbf{R}^2$. It is bounded by these two circles. Thus, for P_3 moving in $\mathcal{R}$, the action variables will change by an amount less than $\mathcal{O}(\mu)$ for all time.

However, if we had been considering a three-dimensional circular restricted problem, the existence of two neighboring invariant KAM tori would not guarantee stability. This is because in this case the phase space is six-dimensional and $\tilde{J}^{-1}(C)$ is five-dimensional. The KAM tori, if they existed by the KAM theorem, would be three-dimensional, as is verified in [204]. Thus, the difference of dimensions is 2, and P_3 would no longer be trapped between them. An analogy is having two concentric circles on a plane in $\mathbf{R}^3$. The instability of motion of a particle P_3 in this higher dimensional situation is called *Arnold diffusion*, [113, 15, 13, 177]. In this case, when P_3 is moving in a resonance region near a three-dimensional torus T^3, it is no longer trapped near it due to geometrical dimensional considerations as we just saw. As shown by Nekhoroshev the rate at which the particle will generally move away from T^3 is very slow. This instability will cause P_3 over a long period of time to drift away from the torus. The first example of this type of instability in another problem was proven by Arnold [13]. It is very difficult to prove that this instability exists in general, and there are few rigorous proofs of its existence.

2.2 THE MOSER TWIST THEOREM, CANTOR SETS

The action, angle variable form of (2.2) with (2.5) can be used in Theorem 2.1 to replace the continuous flow of (2.1) with a two-dimensional area-preserving map on a suitable surface of section $\sum$ of two dimensions which intersects $\mathcal{D}$.

This is done as follows [204]. From (2.5) we can assume $H_{0_{y_1}}(\mathbf{y})$ is bounded away from zero in an open subset $\mathcal{D}^* \subset \mathcal{D}$. Thus, we can solve the equation $H(\mathbf{x}, \mathbf{y}, \mu) = h$ for $y_1 = -\mathcal{F}(\mathbf{x}, y_2, \mu)$, where $\mathcal{F}$ is periodic of period 2π in x_1, x_2, by the implicit function theorem for μ sufficiently small and $\mathcal{D}^*$ sufficiently small. $\mathcal{F}(\mathbf{x}, y_2, 0) = \mathcal{F}(y_2)$ is only a function of y_2. Also,

$$\dot{x}_1 = H_{y_1}(\mathbf{x}, \mathbf{y}) \tag{2.30}$$

is bounded away from zero.

This can be used to eliminate time in (2.1), and we obtain

$$\frac{dx_2}{dx_1} = \frac{H_{y_2}}{H_{y_1}}, \quad \frac{dy_2}{dx_1} = -\frac{H_{x_2}}{H_{y_1}}. \tag{2.31}$$

These differential equations can be written in Hamiltonian form. Consider the identity

$$H(x_1, x_2, -\mathcal{F}(x_1, x_2, y_2, \mu), y_2, \mu) = h.$$

Differentiating with respect to y_2 implies

$$H_{y_2} + H_{y_1}(-\mathcal{F}_{y_2}) = 0$$

or

$$\mathcal{F}_{y_2} = \frac{H_{y_2}}{H_{y_1}}.$$

Similarly, differentiation with respect to x_2 implies

$$\mathcal{F}_{x_2} = \frac{H_{x_2}}{H_{y_1}}. \tag{2.32}$$

Therefore (2.31) implies

$$\frac{dx_2}{dx_1} = \mathcal{F}_{y_2}, \quad \frac{dy_2}{dx_1} = -\mathcal{F}_{x_2}, \tag{2.33}$$

where

$$\mathcal{F} = \mathcal{F}_0(y_2) + \mathcal{F}_1(x_1, x_2, y_2, \mu), \tag{2.34}$$

$\mathcal{F}_1(x_1, x_2, y_2, 0) = 0$. Equation (2.5) implies that $\mathcal{F}_{0y_2y_2} \neq 0$. System (2.33) is a lower dimensional system than (2.1), and is also in action,angle variable form, where $\mathcal{F}_1$ depends also on $\mathbf{x}$. $\mathcal{F}$ is a smooth function of all variables for $y_2 \in \mathcal{D}^*, x_k \in \mathbf{R}^1$, and periodic of period 2π in $x_k, k = 1, 2$.

We now show that (2.33) can be used to construct a two-dimensional area-preserving map on a two-dimensional surface of section $\sum$ to the flow, which by a theorem due to Moser yields invariant closed curves on this section. These curves can be viewed as the intersections of the invariant tori T^2 given in Theorem 1.2 with $\sum$.

Set $\mu = 0$ and let $x_2(0), y_2(0)$ be the values of $x_2(x_1), y_2(x_1)$ at $x_1 = 0$. System (2.33) implies

$$\frac{dx_2}{dx_1} = \mathcal{F}_{0y_2}(y_2), \quad \frac{dy_2}{dx_1} = 0.$$

Thus,

$$y_2 = y_2(0), \quad x_2(x_1) = x_2(0) + x_1\lambda, \quad \lambda = \mathcal{F}_{0_{y_2}}(y_2(0)) = c,$$

$c =$ constant. The two-dimensional map we desire for $\mu = 0$ is given by

$$x_2(2\pi) = x_2(0) + 2\pi\lambda, \qquad y_2(2\pi) = y_2(0), \tag{2.35}$$

where $\lambda = \lambda(y_2(0))$, $d\lambda/dy_2(0) \neq 0, a \leq y_2(0) \leq b, a, b$ are constants.

Equation (2.35) is called a *monotone twist map.* Let's assume $d\lambda/dy_2(0) > 0$. It is defined on the section $\sum$ given by $\{x_1 = 0\}$. It is verified that $\sum$ is a section (or *Poincaré* section) to the flow. This is the case for $\mu = 0$ since $\dot{x}_1 = H_{0_{y_1}}(\mathbf{y}) \neq 0, \mathbf{y} \in \mathcal{D}^*$. That is, the flow intersects the the hyperplane $\{x_1 = 0\}$ transversally.

For notational purposes, we set $y_2(0) \equiv r, y_2(2\pi) \equiv r_1, x_2(0) \equiv \theta, x_2(2\pi) \equiv \theta_1$. Then (2.35) is

$$\theta_1 = \theta + 2\pi\lambda(r), \qquad r_1 = r, \tag{2.36}$$

where $d\lambda/dr \equiv \lambda' > 0, a \leq r \leq b$. We can assume $a \geq 0$, since if a were negative the variable r could be translated, preserving the form of (2.36). This map, M_0, on the annulus $\mathcal{A} = \{0 \leq a \leq r \leq b, 0 \leq \theta \leq 2\pi\}$, is seen to keep each circle invariant by rotating it. The points on a given circle are shifted in a counterclockwise direction by an angle $2\pi\lambda(r)$. r, θ can be viewed as polar coordinates on $\mathcal{A}$. As r increases, the angle increases. The annulus is twisted by M_0. See Figure 2.3.

M_0 is area preserving since (2.33) is a Hamiltonian system, and therefore the flow conserves area elements in the phase space by Liouville's theorem

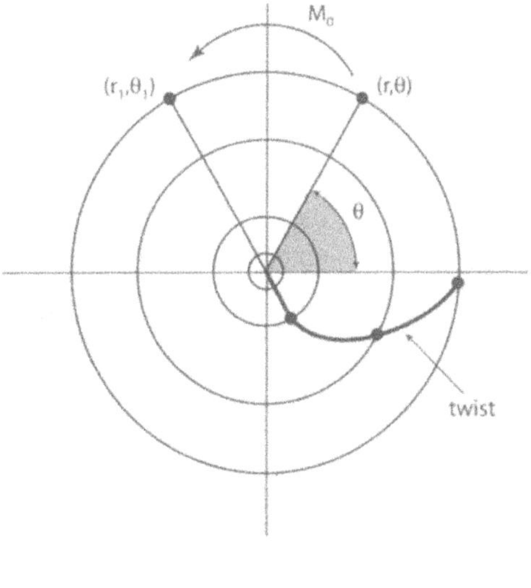

Figure 2.3 Twist map.

[15]. All circles $r = c$ in $\mathcal{A}$ are invariant under M_0, and similar to Theorem 2.1, it would seem likely that those rotation angles $2\pi\lambda(r)$ which are sufficiently irrational would imply that under sufficiently small perturbation μ the corresponding invariant circles are perturbed into invariant closed curves, topologically equivalent to a circle. This is indeed the case, as follows from a fundamental theorem of Moser called the *Moser twist theorem.* Instead of (2.36) we look at its perturbation M_μ due to the term $\mathcal{F}_1$ in the function $\mathcal{F}$ defined by (2.34). More generally, instead of (2.36),

$$\begin{aligned}\theta_1 &= \theta + 2\pi\lambda(r) + f(r,\theta,\mu),\\ r_1 &= r + g(r,\theta,\mu),\end{aligned} \tag{2.37}$$

where $f(r,\theta,0) = g(r,\theta,0) = 0, \lambda' > 0, r \in \mathcal{A}, \lambda(r)$ is smooth, and f, g are smooth in all variables and periodic of period 2π in θ. M_μ is area preserving.

Theorem 2.4 (Moser Twist Theorem) *Assume f, g, λ are $C^j, j \geq 5$, and $|\lambda'(r)| \geq \nu > 0$ on $\mathcal{A}$. Then there exists a $\delta(\mu, \lambda(r)) > 0$ such that if*

$$\sup_{(r,\theta)\in\mathcal{A}} \{\|f\|_j + \|g\|_j\} < \nu\delta, \tag{2.38}$$

then M_μ possesses an invariant curve γ_μ of the form

$$r = c + u(\psi), \quad \theta = \psi + v(\psi) \tag{2.39}$$

in $\mathcal{A}$, where $u, v \in C^1$, are periodic of period 2π, and $\|u\|_1 + \|v\|_1 < \mu, a < c < b$. The induced map on γ_μ is given by

$$\theta \to \theta + 2\pi\lambda, \tag{2.40}$$

where λ is irrational, and satisfies the infinite set of inequalities

$$|\lambda - \frac{n}{m}| \geq \beta m^{-\alpha} \tag{2.41}$$

for some $\beta, \alpha > 0$, and all integers $m, n > 0$. Each choice of λ satisfying (2.41) in the range of $\lambda(r)$ gives rise to such an invariant curve.

This statement of the Moser twist theorem is from [175]. The proof of this theorem is lengthy and requires many delicate estimates. The proof is carried out in [172, 204]. Moser's first proof required that f, g were of class C^{333}, and over subsequent years the requirement on differentiability has gone down with refined proofs. The version of the theorem stated here for class C^5 is due to Rüssman [197]. A counterexample by Takens is given where the differentiability is of class C^1 [220].

The sharpest version of this theorem is due to Herman [102], where $f, g \in C^3$, and so in the Moser twist theorem one can replace C^5 by C^3. For this and other important results pertaining to the existence of invariant closed curves of twist maps, see [102].

Thus as r varies in $\mathcal{A}$, $\lambda(r)$ varies, and when it satisfies (2.41), a perturbed invariant curve is obtained. This is analogous to the situation in Theorem 2.1, where the parameter $\mathbf{c} = (c_1, c_2)$ varied for the tori $T^2(\mathbf{c}, \mu)$ for each fixed μ. We also remark on the notation introduced in Theorem 2.4. C^j represents the class of functions with continuous derivatives up to order $j \geq 0$. Also,

$$\|f\|_j = \sup_{\substack{m+n \leq j \\ \mathcal{A}}} \left| \frac{\partial^{m+n} f}{\partial r^m \partial \theta^n} \right| .$$

We show how to explicitly construct the map (2.37) for (2.21), (2.24), for $\mu = 0$ by the above procedure. The terms f, g in (2.37) are not explicitly constructed but are clearly seen to be smooth functions of μ vanishing at $\mu = 0$ and smooth functions of r, θ, periodic of period 2π in θ, and over the domain for r.

In place of $H_{0_{y_1}}(\mathbf{y})$ bounded away from 0 in the reduction procedure, we have from (2.22)

$$H_{0_{y_2}} = -1.$$

This defines a surface of section $\sum = \{x_2 = 0\}$ with coordinates x_1, y_1, implying by the implicit finction theorem that $y_2 = -\mathcal{F}(\mathbf{x}, y_1, \mu)$ and $\dot{x}_2 \neq 0, x_2 = x_2(0) = 0$. Thus, y_2 can be eliminated and x_2 eliminated and used for the independent variable. Equation (2.33) is replaced by

$$\frac{dx_1}{dx_2} = -H_{y_1} = y_1^{-3}, \qquad \frac{dy_1}{dx_2} = 0$$

which implies $y_1(2\pi) = y_1(0)$,

$$x_1(2\pi) = x_1(0) - \int_0^{2\pi} y_1^{-3}(0) dx_2 = x_1(0) - 2\pi y_1^{-3}(0).$$

Setting $\lambda = -y_1^{-3}(0), d\lambda/dy_1(0) = 3y_1^{-4}(0) \neq 0$ yields (2.35). The resulting map $x_1(2\pi) = x_1(0) + 2\pi\lambda(y_1(0))$, $y_1(2\pi) = y_1(0)$, is defined on the section $\sum$. Thus, returning to (2.33), Theorem 2.4 guarantees the existence of the invariant curves on the section $\sum$ which intersects $\mathcal{D}^* \in H^{-1}(h)$. This set $\mathcal{C}$ of invariant curves $\gamma_\mu(r)$, is topologically equivalent to circles and can be viewed as the intersection of the KAM tori of Theorem 2.3 with $\sum$. (See Figure 2.4.) It is proven in [204] that the set $\mathcal{C}$ is a set of positive measure, and the set Λ of the excluded points on $\sum$ not satisfying (2.41) is small so that a majority of the points belong to $\mathcal{C}$. Λ is called the set of *resonance gaps* or the *resonance zone* for the map.

2.2.1 Cantor Sets

Examination of the set of ω satisfying (2.41) shows that this set is totally disconnected, and it is perfect. It is a Cantor set. It is recalled that a set

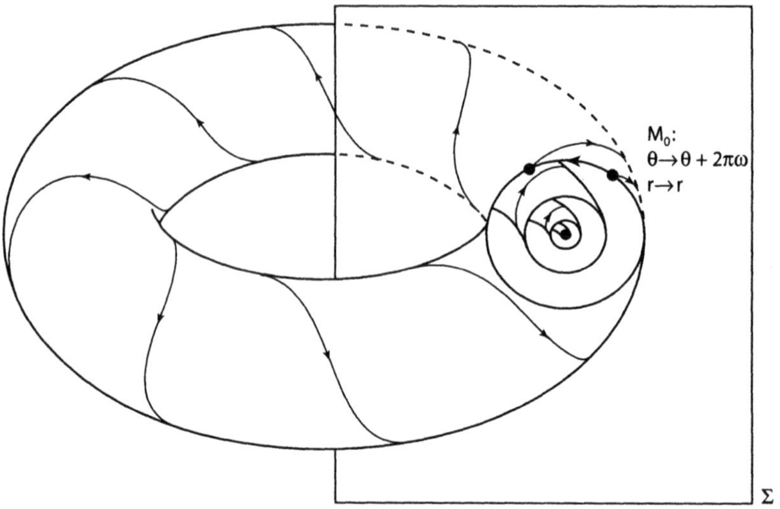

Figure 2.4 Twist map induced on a section of KAM tori.

$E \subset \mathbf{R^1}$ is *totally disconnected* if given any two points $x, y \in E$, there exist separated sets $A, B \subset \mathbf{R^1}$ with $x \in A, y \in B$, and $E = A \cup B$. A and B are separated if there exist disjoint open sets U, V with $A \subseteq U, B \subseteq V$. A *perfect set* $P \subseteq \mathbf{R^1}$ is closed and contains no isolated points. An isolated point of a set $D \subseteq \mathbf{R^1}$ is a point x where one can find an open neighborhood N such that $N \cap D = \{x\}$.

Definition 2.5 *A Cantor set in* $\mathbf{R^1}$ *is a set that is totally disconnected, perfect, and bounded. The following theorem is proven in [1].*

Theorem 2.6 *A nonempty perfect set in* $\mathbf{R^1}$ *is uncountable.*

Thus, we have the following lemma.

Lemma 2.7 *A Cantor set is uncountable and totally disconnected.*

A classic example of a Cantor set in $\mathbf{R^1}$ is obtained by the following construction due to Georg Cantor. Let C_0 be the closed interval $[0, 1]$. Let C_1 be the set obtained from removing the middle third of C_0, $C_1 = [0, 1/3] \cup [2/3, 1]$. Continue by removing the middle third of the two intervals composing C_1 so that $C_2 = ([0, 1/9] \cup [2/9, 1/3]) \cup ([2/3, 7/9], [8/9, 1])$. Inductively continue this process to construct a sequence of sets $C_n, n = 1, 2, 3, \ldots$, where C_n has 2^n closed intervals each of length $1/3^n$. Set

$$C = \cap_{n=0}^{\infty} C_n.$$

It would seem that C is a very sparse set since so many intervals are removed, as is seen in Figure 2.5.

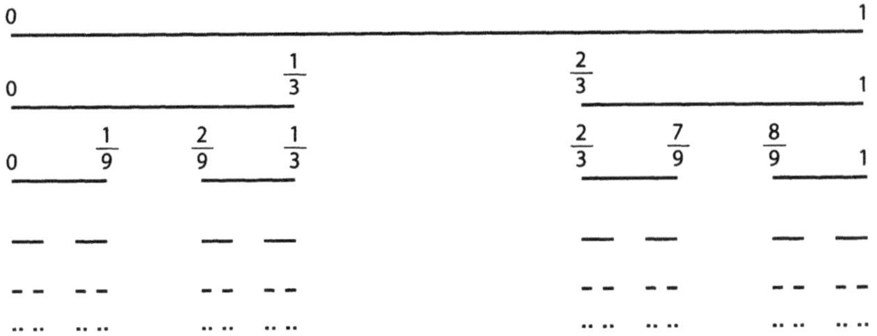

Figure 2.5 Classic example of construction of a Cantor set.

One refers to C as being like *dust* on the line; this is sometimes referred to as *Cantor dust.* For reference, we refer to C as the *Cantor discontinuum.* Nevertheless, Cantor proved the following theorem.

Theorem 2.8 *C is a totally disconnected perfect set and therefore uncountable with cardinality equal to $\mathbf{R}^1$, and $m(C) = 0$.*

This may seem nonintuitive. C has the same cardinality as $\mathbf{R}^1$ (i.e., C can be put in a $1:1$ correspondence with $\mathbf{R}^1$), but yet the total length of all the points when pushed together is 0. That is, C has the same length as a point. It is possible to construct a Cantor set on $\mathbf{R}^1$ with positive measure. This is done by the following construction. Instead of removing the middle one-third intervals as in C, we instead remove one-third intervals while magnifying C by a factor of 3. In other words, C_0 is magnified to $[0, 3]$, and then $C_1 = [0, 1] \cup [2, 3]$. Continuing in this way, it is seen that magnifying the Cantor set by 3 yields another copy of C_0. Thus, if the resulting set is called C, then we must satisfy for the length, or dimension of C, $2 = 3^x$, which implies

$$x = \frac{\ln 2}{\ln 3} \cong 0.631,$$

yielding a fractional dimension.

Cantor sets of positive measure on the line therefore can have fractional dimension in general. This is related the fact that a Cantor set on $\mathbf{R}^1$ has an intricate structure.

Cantor sets are more generally defined in [108] for a metric space of finite dimension.

Definition 2.9 *A* Cantor set *is a totally disconnected, perfect, compact metric space.*

The following topological classification theorem is proven in [108].

Theorem 2.10 *Any two totally disconnected, perfect, compact metric spaces are homeomorphic.*

Thus, the Cantor discontinuum can be used as the standard model for any Cantor set. Theorem 2.10 implies the following corollary.

Corollary 2.11 *Any Cantor set is homeomorphic to the Cantor discontinuum.*

2.3 AREA-PRESERVING MAPS, FIXED POINTS, HYPERBOLICITY

Cantor sets are associated to a particular type of complicated dynamics. This is the case for the dynamics in the set Λ for the general map M_μ defined by (2.37). Before this is described, we make some basic definitions.

Instead of M_μ, we consider a general area-preserving map $\mathcal{M} : (x, y) \to (x_1, y_1)$,

$$x_1 = f(x, y), \qquad y_1 = g(x, y) \tag{2.42}$$

of the (x, y)-plane where f, g are smooth in a neighborhood N of $x = 0, y = 0$, which is a fixed point. That is, $f(0,0) = g(0,0) = 0$, so that $(x, y) = (0,0)$ is mapped into itself. We use the following notation: $\mathbf{X}_1 = \mathcal{M}(\mathbf{X})$, $\mathbf{X} = (x, y)$, $\mathbf{X}_1 = (x_1, y_1)$, $\mathcal{M}^2(\mathbf{X}) = \mathcal{M}(\mathcal{M}(\mathbf{X}))$, $\mathcal{M}^0 = I, \mathcal{M}^{-1}$ is the inverse map, I is the identity map. The origin is chosen without loss of generality since otherwise a point $(x_0, y_0) \neq (0,0)$ can be shifted to the origin by a constant translation which will preserve the form of (2.42). The linearization of this map at the origin is given by

$$\mathbf{X}_1 = \mathbf{A}\mathbf{X}, \tag{2.43}$$

$\mathbf{X} \in U \subset N$,

$$\mathbf{A} = D_{\mathbf{X}}\mathbf{F}(\mathbf{0}) \equiv \frac{\partial \mathbf{F}(\mathbf{X})}{\partial \mathbf{X}} |_{\mathbf{X}=\mathbf{0}} = \left(\begin{array}{cc} f_x & f_y \\ g_x & g_y \end{array} \right) \Bigg|_{\mathbf{X}=\mathbf{0}},$$

where U is a neighborhood of $\mathbf{0}$, and $\mathbf{F}(\mathbf{X}) = (f(\mathbf{X}), g(\mathbf{X}))$.

$\mathbf{A^2X} \equiv \mathbf{A}(\mathbf{AX})$ represents two applications of $\mathbf{A}$. Continuing in this way, one defines n applications of $\mathbf{A}$, $\mathbf{A^nX}$, $n = 1, 2, 3, \ldots$. Equation (2.43) is just the first term in the Taylor series expansion of $\mathbf{F}$ about $\mathbf{X} = \mathbf{0}$. $\mathbf{A} = D_{\mathbf{X}}\mathbf{F}(\mathbf{0}) \equiv (a_{ij})$ is a constant matrix. The fact that (2.42) is area-preserving implies $f_x g_y - f_y g_x = 1$, and for the linearization (2.43), $a_{11}a_{22} - a_{12}a_{21} = 1$. This implies that the matrix $\mathbf{A}$ is invertible and has eigenvalues $\lambda_1, \lambda_2 \in \mathbb{C}$. They satisfy the equation

$$|\mathbf{A} - \lambda \mathbf{I}| = \lambda^2 - (a_{11} + a_{22})\lambda + a_{11}a_{22} - a_{12}a_{21} = 0,$$

where $\mathbf{I}$ is the 2×2 identity matrix.

Since (2.42) is area-preserving, this reduces to

$$\lambda^2 - (a_{11} + a_{22})\lambda + 1 = 0.$$

λ_1, λ_2 satisfy $\lambda_1 + \lambda_2 = a_{11} + a_{22}$ and $\lambda_1\lambda_2 = 1$. We have three cases: (i) hyperbolic, where $\lambda_1 \neq \lambda_2$ and $\lambda_i \in \mathbf{R^1}$; (ii) elliptic, where $\lambda_1 = \bar{\lambda}_2, \lambda_1 \neq \lambda_2$; (iii) $\lambda_1 = \lambda_2$.

We exclude case (iii) and only consider cases (i) and (ii).

Definition 2.12 *In case (i),* $(x, y) = (0, 0)$ *is called a* hyperbolic fixed point *for (2.42), and for case (ii),* $(x, y) = (0, 0)$ *is called an* elliptic fixed point *for (2.42).*

Let's consider case (i) first. In a neighborhood of $(0, 0)$, $\mathbf{R^2}$ decomposes into two one-dimensional subspaces

$$\mathbf{R^2} = \mathbf{E}^+ \oplus \mathbf{E}^-,$$

where $\mathbf{E}^+, \mathbf{E}^-$ are *invariant subspaces* under $\mathbf{A}$ and are each one-dimensional. This means that if $\mathbf{u} \in \mathbf{E}^+, \mathbf{v} \in \mathbf{E}^-, \mathbf{u} \neq 0, \mathbf{v} \neq 0$, then $\mathbf{A^nu} \in \mathbf{E}^+, \mathbf{A^nv} \in \mathbf{E}^-, n = \pm 1, \pm 2, \pm 3, \ldots$. The set $\{\mathbf{A^nu}, n = \pm 1, \pm 2, \ldots\}$ represents the *trajectory* of $\mathbf{u}$ and is a discrete set of points. This definition of a trajectory is analogous to that of a trajectory of a differential equation, which depends continuously on time.

$\mathbf{E}^+, \mathbf{E}^-$ are the eigenspaces spanned by the eigenvectors $\mathbf{e}_+, \mathbf{e}_-$, respectively, associated to the eigenvalues λ_1, λ_2, respectively, of $\mathbf{A}$, where we assume $|\lambda_1| < 1, |\lambda_2| > 1$, since $\lambda_1\lambda_2 = 1$.

Definition 2.13 $\mathbf{E}^+(\mathbf{0}), \mathbf{E}^-(\mathbf{0})$ *are one-dimensional invariant manifolds for* $\mathbf{A}$ *satisfying*

$$\mathbf{E}^+(\mathbf{0}) = \{\mathbf{x} \in U | \mathbf{A^n x} \to \mathbf{0}, n \to \infty, \mathbf{A^n x} \in U, n = 0, 1, \ldots\},$$
$$\mathbf{E}^-(\mathbf{0}) = \{\mathbf{x} \in U | \mathbf{A^{-n} x} \to \mathbf{0}, n \to \infty, \mathbf{A^{-n} x} \in U, n = 0, 1, \ldots, \}.$$

$\mathbf{E}^+(\mathbf{E}^-)$ *are called* invariant stable (unstable) hyperbolic manifolds.

Definition 2.14 $\mathbf{A}$ *restricted to* $\mathbf{E}^+, \mathbf{E}^-$ *is called a* contraction, expansion, *respectively, since*

$$|\mathbf{Au}| = |\lambda_1 \mathbf{u}| < |\mathbf{u}|, \quad |\mathbf{Av}| = |\lambda_2 \mathbf{v}| > |\mathbf{v}|.$$

The dynamics of $\mathbf{A}$ is illustrated in Figure 2.6, where the directions indicated refer to the directions of the iterates $\mathbf{A^n x}, n = 1, 2, \ldots, \mathbf{x} \in \mathbf{U}$. This is called a *hyperbolic saddle.*

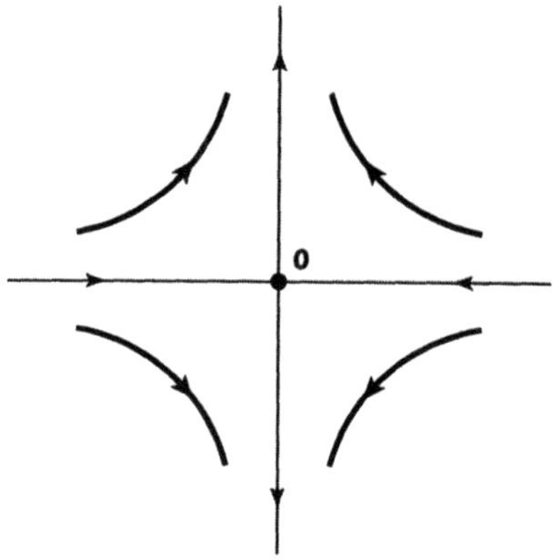

Figure 2.6 Hyperbolic point.

For (x, y) sufficiently near the origin, or equivalently for U sufficiently small, $\mathcal{M}$ is uniformly close to $\mathbf{A}$, and the dynamics of $\mathcal{M}$ turns out to very close to that of $\mathbf{A}$.

Theorem 2.15 *There exist local smooth one-dimensional stable and unstable manifolds* $\mathbf{W^s}(\mathbf{0}), \mathbf{W^u}(\mathbf{0})$ *in* U *for* $\mathcal{M}$, *tangent at* $\mathbf{X} = \mathbf{0}$ *to the invariant manifolds* $\mathbf{E}^+, \mathbf{E}^-$, *respectively, for* $\mathbf{A}$, *where*

$$\mathbf{W^s}(\mathbf{0}) = \{\mathbf{x} \in U | \mathcal{M}^n(\mathbf{x}) \to \mathbf{0} \text{ as } n \to \infty, \mathcal{M}^n(\mathbf{x}) \in U, n = 0, 1, \ldots, \},$$
$$\mathbf{W^u}(\mathbf{0}) = \{\mathbf{x} \in U | \mathcal{M}^{-n}(\mathbf{x}) \to \mathbf{0} \text{ as } n \to \infty, \mathcal{M}^{-n}(\mathbf{x}) \in U, n = 0, 1, \ldots\}.$$

Theorem 2.15 can be viewed as a special case of the *stable manifold theorem*, which more generally requires that $\mathcal{M}$ be a C^k diffeomorphism, $k = 1, 2, \ldots$ (i.e., a C^k map with a C^k inverse map). C^∞ means $\mathcal{M}$ has continuous derivatives of all orders, and C^Ω means $\mathcal{M}$ is real analytic (smooth). In this more general setting $\mathcal{M}$ need not be area-preserving, and the condition on being a hyperbolic fixed point is replaced with $\mathbf{A}$ having no eigenvalues of unit modulus. The stable manifold theorem is valid for n dimensions and where $\mathbf{W^u}, \mathbf{W^s}$ have the same degree of differentiability as $\mathcal{M}$. (See [89, 91].)

In [204] the following theorem due to Birkhoff is proven for $\mathcal{M}$ in the hyperbolic case.

Theorem 2.16 *There exists a smooth change of coordinates $(x, y) \to (\xi, \eta)$ in the hyperbolic case for N sufficiently small where $\mathcal{M}$ has the* normal form

$$\xi_1 = \pm e^{w}\xi, \qquad \eta_1 = \pm e^{-w}\eta, \tag{2.44}$$

where $w = \gamma_0 + (\xi\eta)^k, \lambda_1 = \pm e^{\gamma_0}, k \geq 1$, or $w = \gamma_0, \gamma_0 \neq 0$.

An analogous result for the elliptic case is given by the following theorem.

Theorem 2.17 *There exists a smooth canonical change of coordinates $(x, y) \to (\xi, \eta)$ in the elliptic case for N sufficiently small where $\mathcal{M}$ can be written in the normal form*

$$\begin{aligned} \xi_1 &= \xi \cos w_k - \eta \sin w_k + \mathcal{O}_{2k+2}, \\ \eta_1 &= \xi \sin w_k + \eta \cos w_k + \mathcal{O}_{2k+2}, \\ w_k &= \gamma_0 + \gamma_k(\xi^2 + \eta^2)^k, \end{aligned} \tag{2.45}$$

where it is assumed that $\lambda_1^j \neq 1, j = 1, \ldots, 2k+2$, $\gamma_i = 0, i = 1, \ldots, k-1, \gamma_k \neq 0$, and where γ_0 is a constant, $\gamma_k > 0, k > 0$, and the terms $\mathcal{O}_{2k+2}$ are power series in ξ, η containing terms of order $\geq 2k+2$ only.

Equation (2.45) without the form $\mathcal{O}_{2k+2}$ is a rotation by the angle w_k. This theorem is proven in [204]. A version of this was proven by Birkhoff [46], where convergence properties of the power series representation was not considered. The proof of the theorem in [204] avoids convergence difficulties by truncation of some power series representations. Theorem 2.17 assumes $\mathcal{M}$ is smooth; i.e., real analytic. In [175], it is only necessary to assume that $\mathcal{M}$ is $C^\ell, \ell \geq 4$, where $\lambda_1^k \neq 1$ for $k = 1, 2, \ldots, q, 4 \leq q \leq \ell + 1$.

It will turn out, as we will see later in the section in Lemma 2.20, that (2.45) can be written in the form of the twist map (2.37) which will be used to prove that the elliptic fixed point $(x, y) = (\xi, \eta) = (0, 0)$ is *stable*.

So far we have considered the *local* existence of $\mathbf{W^u}, \mathbf{W^s}$ for $\mathcal{M}$ near $(0,0)$. We now consider these manifolds *globally*, where in general they will extend beyond N. Let $\mathbf{W}^{\mathbf{u}}_{\text{loc}}(\mathbf{0}), \mathbf{W}^{\mathbf{s}}_{\text{loc}}(\mathbf{0})$ denote the local manifolds $\mathbf{W^u}(\mathbf{0}), \mathbf{W^s}(\mathbf{0})$ given by Theorem 2.15. We define the global manifolds

$$\mathbf{W^s}(\mathbf{0}) = \cup_{n\geq 0}\mathcal{M}^{-n}(\mathbf{W}^{\mathbf{s}}_{\text{loc}}(\mathbf{0})),$$
$$\mathbf{W^u}(\mathbf{0}) = \cup_{n\geq 0}\mathcal{M}^{n}(\mathbf{W}^{\mathbf{u}}_{\text{loc}}(\mathbf{0})).$$

For this discussion, let $\mathbf{p}$ be a hyperbolic fixed point of $\mathcal{M}$, and let $\mathbf{q}$ be another hyperbolic fixed point of $\mathcal{M}, \mathbf{p} \neq \mathbf{q}$. We will consider two cases. The first case is shown in Figure 2.7. $\mathbf{W^u}(\mathbf{p})$ coincides with $\mathbf{W^s}(\mathbf{p})$.

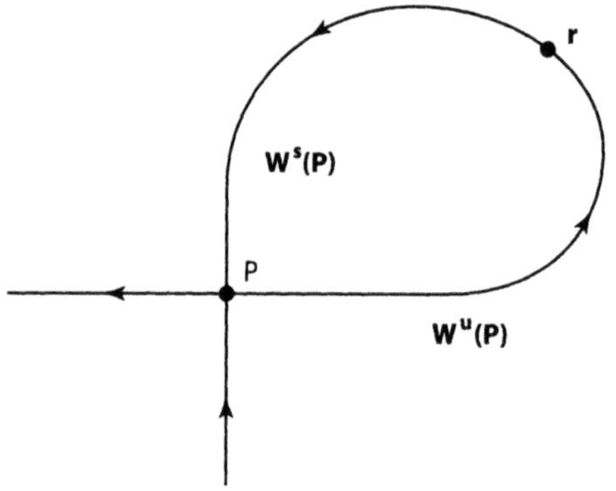

Figure 2.7 Homoclinic point $\mathbf{r}$.

Definition 2.18 $\mathbf{x}$ *is called a* homoclinic point *if for any point* $\mathbf{x} \in \mathbf{W^u}(\mathbf{p}) \cap \mathbf{W^s}(\mathbf{p}), \mathbf{x} \neq \mathbf{p}, \mathcal{M}^n(\mathbf{x}) \to \mathbf{p}, \mathcal{M}^{-n}(\mathbf{x}) \to \mathbf{p}, n \to \infty$.

The example in Figure 2.7 occurs when $\mathbf{W^u}(\mathbf{p}), \mathbf{W^s}(\mathbf{p})$ coincide at $\mathbf{x}$. The resulting invariant manifold which escapes $\mathbf{p}$ on $\mathbf{W^u}(\mathbf{p})$ and approaches $\mathbf{p}$ again on $\mathbf{W^s}(\mathbf{p})$ is called a *homoclinic loop* [89]. In Figure 2.7 $\mathbf{W^s}, \mathbf{W^u}$ coincide. However, this need not be the case, as they can intersect transversally at a point $\mathbf{r} \in \mathbf{W^s}(\mathbf{p}) \cap \mathbf{W^u}(\mathbf{p})$. This means that the tangent spaces to $\mathbf{W^s}(\mathbf{p}), \mathbf{W^u}(\mathbf{p})$ do not coincide. $\mathbf{r}$ is then called a *transverse homoclinic point.*

When this happens the resulting dynamics $\mathcal{M}$ gets very complicated. The curve $\mathbf{W^u}(\mathbf{p})$ snakes (or winds) its way under forward iterates for the points on it towards $\mathbf{p}$ in the manner shown in Figure 2.8. That is, it intersects $\mathbf{W^s}(\mathbf{p})$ infinitely many times. This must be the case since for any $\mathbf{q} \in \mathbf{W^s}, \mathcal{M}^n(\mathbf{q}) \to \mathbf{p}$ as $n \to \infty$. As $\mathbf{W^u}(\mathbf{p})$ gets pushed towards $\mathbf{p}$, the loop arcs must get stretched and become thinner since $\mathcal{M}$ must preserve area in the interiors of each arc area enclosed by the $\mathbf{W^u}$ and $\mathbf{W^s}$ as they get mapped under forward iterations. The same thing is seen as $\mathbf{W^s}$ gets pushed

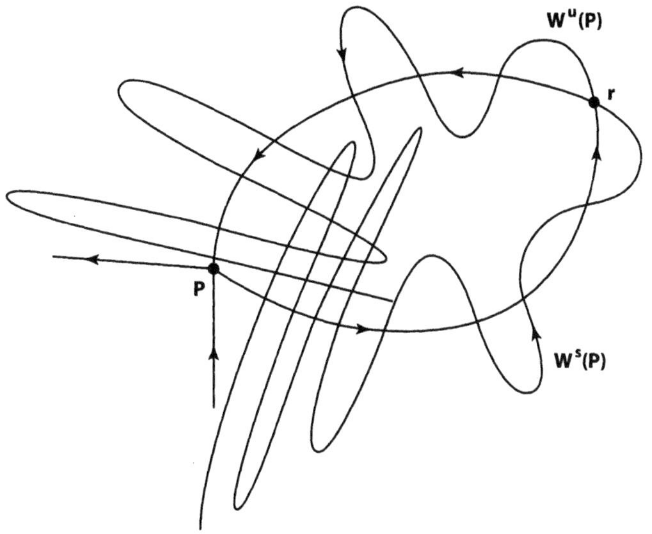

Figure 2.8 Transverse homoclinic point **r** and hyperbolic network.

towards **p** under backwards iterates for the points on it. As $\mathbf{W^u(p)}$, for example, snakes towards **p** in forward iterations it cannot cross the portion of $\mathbf{W^u}$ moving away from **p** by uniqueness of solutions, and likewise $\mathbf{W^s(p)}$ for backwards iterates will not cross itself. However, $\mathbf{W^u}, \mathbf{W^s}$ can intersect [89]. Infinitely many intersecting points of $\mathbf{W^s} \cap \mathbf{W^u}$ appear, of which all are homoclinic points. An extremely complicated mesh of intersection points eventually occurs. In the limit, this set is dense. This mesh or *hyperbolic network* is discussed further in chapter 3 in terms of its structure, which turns out to be a Cantor set.

It is clear that the fate of a point chosen in the hyperbolic network is extremely complicated. Tiny changes in the location of the initial point will lead, in general, to a very different trajectory of iterates. The existence of a network formed in this way leads to a very sensitive, or *chaotic* dynamics for $\mathcal{M}$. This is the way the term *chaotic* is used in this book. That is, the dynamics of points on a set S under a map $\mathcal{M}$ is chaotic if it can be shown to be equivalent to a map on a hyperbolic network formed from a transverse homoclinic or heteroclinic point, defined below. More generally, we say that the dynamics of points on the set S are chaotic if S is a *hyperbolic invariant set* defined in section 3.6.2.

This, of course, is a very restrictive definition of chaos, as complicated sensitive motions can occur in many ways, like a leaf being blown about in the wind. However, for this book, the definition just given is suitable.

Another situation we consider is shown in Figure 2.9. There we consider two fixed hyperbolic points of $\mathcal{M}, \mathbf{p} \neq \mathbf{q}$.

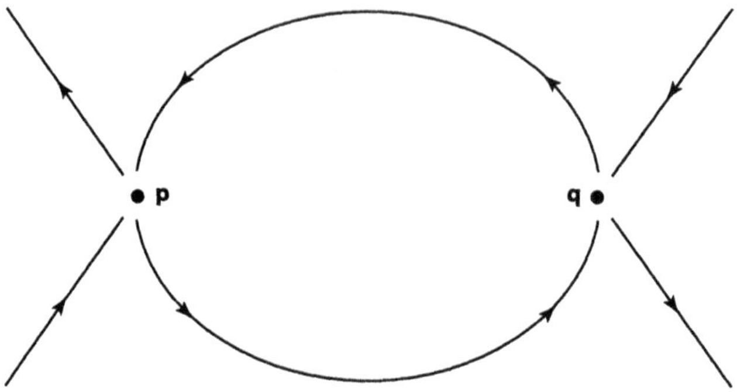

Figure 2.9 Heteroclinic connection.

Definition 2.19 $\mathbf{x}$ *is a* heteroclinic point *for* $\mathcal{M}$ *if* $\mathbf{x} \in \mathbf{W^u}(\mathbf{p}) \cap \mathbf{W^s}(\mathbf{q}) - \{\mathbf{p}, \mathbf{q}\}$.

When $\mathbf{W^u}(\mathbf{p}), \mathbf{W^s}(\mathbf{p})$ *do not* intersect transversally as shown in Figure 2.9, the connecting manifolds between $\mathbf{p}, \mathbf{q}$ comprise a *heteroclinic connection.* If $\mathbf{W^u}(\mathbf{p}), \mathbf{W^s}(\mathbf{q})$ do intersect transversally at a point $\mathbf{r}$, the same type of situation discussed already for the breaking of a homoclinic loop occurs, infinitely many heteroclinic points are formed, and a complicated hyperbolic network is created as shown in Figure 2.10. In this case, $\mathbf{r}$ is called a *transverse heteroclinic point.*

Let us now regard $\mathcal{M}$ as being defined as a map on section σ of two dimensions to a flow of a system S of differential equations, say a system of two first order differential equations in two variables explicitly depending on the time variable with periodic dependence. This is analogous to (2.33) which we used to construct the map M_μ on the section $\sum$. A fixed point of $\mathcal{M}$ on σ defines a periodic orbit. This is seen in Figure 2.11 for the hyperbolic fixed point $\mathbf{0}$ for $\mathcal{M}$ with local manifolds $\mathbf{W^u_{loc}}(\mathbf{0}), \mathbf{W^s_{loc}}(\mathbf{0})$. The periodic orbit $\phi(t)$ itself and the flow near it is shown.

In the three-dimensional extended phase space in Figure 2.11, the one-dimensional invariant manifolds $\mathbf{W^u}, \mathbf{W^s}$ for $\mathcal{M}$ become two-dimensional invariant manifolds $\mathbf{W^u_{loc}}(\phi), \mathbf{W^s_{loc}}(\phi)$ for the flow of S. The flow moves towards ϕ on $\mathbf{W^s_{loc}}(\phi)$ and away from ϕ on $\mathbf{W^u_{loc}}(\phi)$.

In Figure 2.12 we show what the global invariant manifolds for a system S would look like near ϕ and σ. This figure shows Figure 2.8 in the extended phase space. In this case $\mathbf{W^u}(\phi), \mathbf{W^s}(\phi)$ intersect transversally at

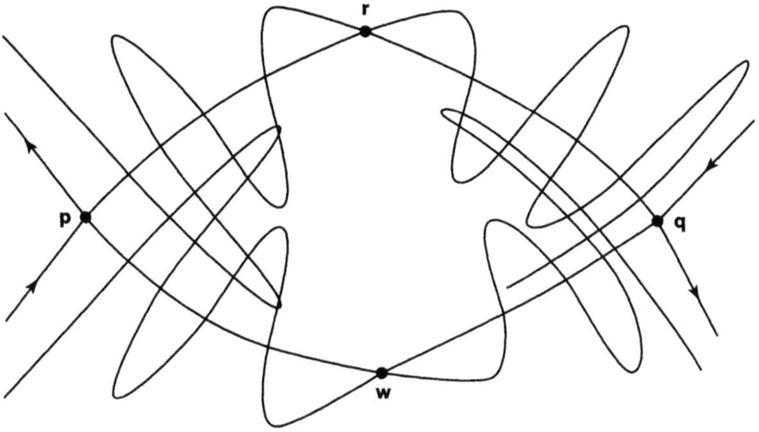

Figure 2.10 Transverse heteroclinic points $\mathbf{r}, \mathbf{w}$ and hyperbolic network.

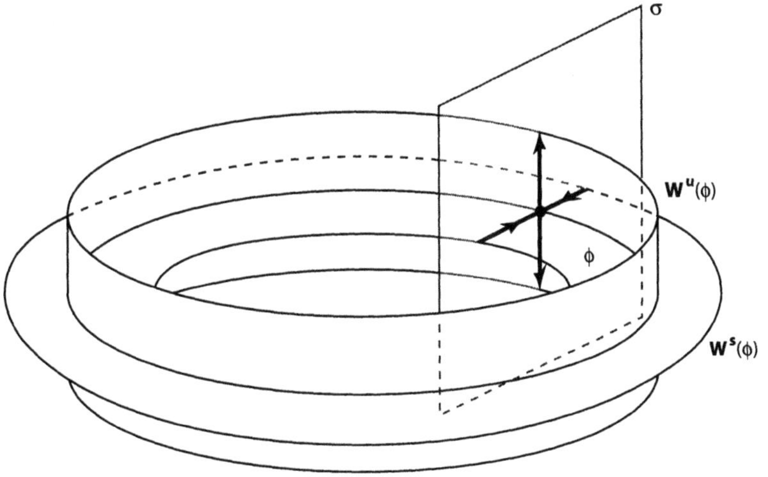

Figure 2.11 Section of a hyperbolic periodic orbit.

a homoclinic point on σ. This point is a transverse homoclinic point of the orbit.

It is important to note that the intersections of $\tilde{\mathbf{W}}^{\mathbf{u}} \equiv \mathbf{W}^{\mathbf{u}}(\phi), \tilde{\mathbf{W}}^{\mathbf{s}} \equiv \mathbf{W}^{\mathbf{s}}(\phi)$ in Figure 2.12 do not violate uniqueness theorems in the theory of ordinary differential equations. This is because the intersecting manifolds are $\tilde{\mathbf{W}}^{\mathbf{u}}, \tilde{\mathbf{W}}^{\mathbf{s}}$, and not $\tilde{\mathbf{W}}^{\mathbf{s}}, \tilde{\mathbf{W}}^{\mathbf{s}}$ or $\tilde{\mathbf{W}}^{\mathbf{u}}, \tilde{\mathbf{W}}^{\mathbf{u}}$. We consider a solution curve $\psi(t)$ for $-\infty < t < \infty$. We assume $\psi \in \tilde{\mathbf{W}}^{\mathbf{u}}$, so that for t near $-\infty, \psi(t)$ is arbitrarily close to $\phi(t)$. As t increases, ψ winds its way up $\tilde{\mathbf{W}}^{\mathbf{u}}$, moving chaotically with many oscillations. It stays on $\tilde{\mathbf{W}}^{\mathbf{u}}$, passes through many

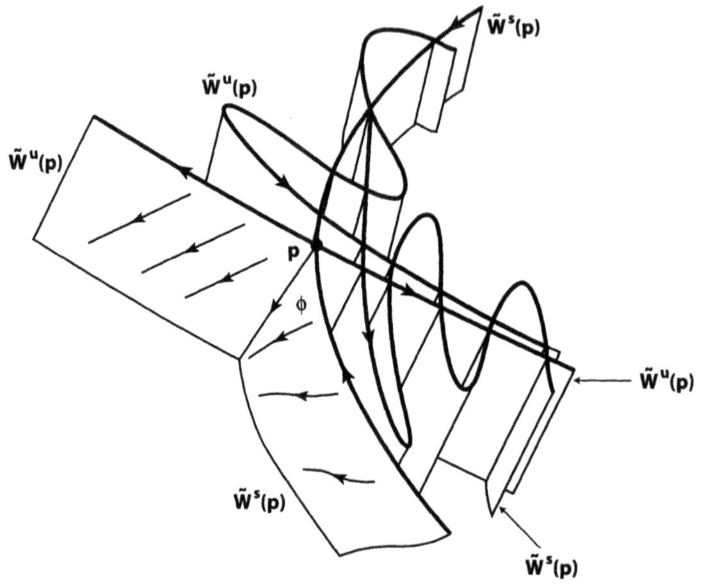

Figure 2.12 Intersections of invariant manifolds near a hyperbolic orbit.

points of intersection with $\tilde{\mathbf{W}}^s$, and intersects no other solutions at those intersection times. Thus, no uniqueness theorems are violated since the different solutions on the invariant manifolds do not intersect one another, even though the manifolds themselves intersect. One can imagine an infinitely long surface of infinitely many self-intersections stretching from $t = -\infty$ to $t = +\infty$. A cross section to it on σ would appear as in Figure 2.8, and near to ϕ would appear as in Figure 2.12. Solutions move up this manifold as t increases, moving through all the intersections. This situation is described in section 3.6. In Figure 2.13 the homoclinic loop on σ does not break and a cross section to it would appear as in Figure 2.7. The sections are viewed at every time $t = T$, T is the period of ϕ. One can see a point moving on the homoclinic loop for the orbit $\varphi(t)$ on σ, labeled $1, 2, 3, 4, 5$.

We now return to the map $\mathcal{M}$ defined by (2.42) in the case of an elliptic fixed point $(0, 0)$. We assume that it satisfies the assumptions of Theorem 2.17 and thus can be written in the form (2.45). Under this assumption, we can state the following key lemma.

Lemma 2.20 *The area-preserving map $\mathcal{M}$ of the elliptic type represented by (2.45) can be written in the form of the twist map M_μ given by (2.37).*

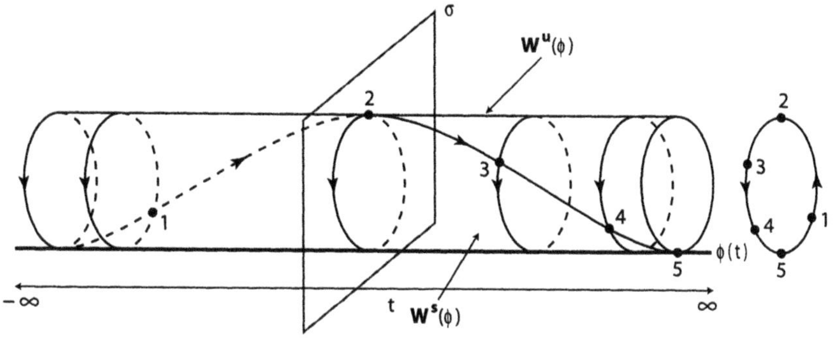

Figure 2.13 Orbit moving from stable to unstable manifolds in extended phase space.

Proof. Consider (2.45). Transform to polar coordinates

$$\xi = \tilde{\mu} r^{\frac{1}{2k}} \cos\theta, \quad \eta = \tilde{\mu} r^{\frac{1}{2k}} \sin\theta, \tag{2.46}$$

$\tilde{\mu} > 0$, yielding the transformed map

$$\theta_1 = \theta + 2\pi\lambda(r) + \mathcal{O}(\tilde{\mu}^{2k+1}), \qquad r_1 = r + \mathcal{O}(\tilde{\mu}^{2k+1}), \tag{2.47}$$

as is verified. The error terms are smooth in r, θ and periodic of period 2π in θ. We assume $\xi^2 + \eta^2 < \tilde{\mu}^2$, which implies $0 < r < 1$. This is in the form of (2.37), where $\lambda(r) = (2\pi)^{-1}(\gamma_0 + \gamma_k \tilde{\mu}^{2k} r)$, implying $d\lambda/dr = (2\pi)^{-1}\gamma_k \tilde{\mu}^{2k} > 0$. □

Lemma 2.21 *Equation (2.45) is stable at the fixed point* $(0,0)$.

Proof. In (2.47) we restrict r to a closed subinterval of $(0,1)$. This is required since r in Theorem 2.4 belongs to the closed annulus $\mathcal{A}$.

Theorem 2.4 implies that there exists a δ such that

$$\frac{2\pi}{\gamma_k \tilde{\mu}^{2k}} \sup_{(r,\theta)\in\mathcal{A}} \{\|f\|_j + \|g\|_j\} < \delta,$$

where in our case $f, g = \mathcal{O}(\tilde{\mu}^{2k+1})$. This estimates the error terms f, g which imply they are of magnitude $\mathcal{O}(\tilde{\mu})$, and for $\tilde{\mu}$ sufficiently small, they can be made arbitrarily small.

Thus, by Theorem 2.4, for each $\tilde{\mu}$ sufficiently small, in the punctured disc $0 < \xi^2 + \eta^2 < \tilde{\mu}^2$ there exists an invariant curve Γ (and actually infinitely many of them satisfying condition (2.41)). The iterates of $M_{\tilde{\mu}}$ and hence $\mathcal{M}$ are therefore trapped within the region enclosed by Γ. Thus, $\mathcal{M}$ is stable at the fixed point $(0,0)$. It is noted that the parameter $\tilde{\mu}$ in $M_{\tilde{\mu}}$ representing the distance to the origin is not μ, which is from the restricted problem related to the mass ratio between m_1, m_2. □

Note that it is necessary to assume that the terms $\mathcal{O}_{2k+2}$ in Theorem 2.17 do not identically vanish. Otherwise counterexamples exist where area-preserving maps near an elliptic fixed point are not stable in the case where (2.42) is of class C^k, k arbitrary [120].

2.3.1 Geometry Near the Fixed Point

Thus, we have a family of invariant curves $\Gamma(\tilde{\mu})$, where $\tilde{\mu} > 0$ is sufficiently small. In the (ξ, η)-space they are in the punctured disc $0 < \xi^2 + \eta^2 < \tilde{\mu}^2$ and enclose the origin. They converge to the origin as $\tilde{\mu} \to 0$. Hence, *under the assumption of Theorem 2.17* there are infinitely many invariant curves for $\mathcal{M}$, and they exist provided condition (2.41) is satisfied.

As previously discussed for M_μ, this yields a Cantor set $\tilde{\mathcal{C}}$ of positive measure of invariant closed curves, where the complement $\tilde{\Lambda}$ of $\tilde{\mathcal{C}}$ in $0 < \xi^2 + \eta^2 < \rho^2 < \tilde{\mu}^2$ has small measure $o(\pi\rho^2)$.

The set $\tilde{\mathcal{C}}$ is totally disconnected and uncountable as follows from (2.41). Curves in $\tilde{\mathcal{C}}$ can be abstractly pictured as *rings of a tree*. The gaps between the curves represent the set $\tilde{\Lambda}$. The gaps occur when the rotation number ω is sufficiently close to being a rational number.

Note that the general map $\mathcal{M}$ given by (2.42) satisfying the assumptions of Theorem 2.17 is by Lemma 2.20 equivalent to the twist map M_μ, (2.37), $\mu \equiv \tilde{\mu}$. So, without loss of generality, we can consider twist maps M_μ of the form (2.37) satisfying Theorem 2.4.

Note that the particular map M_μ constructed in section 2 for the restricted problem on $\sum$ is of the form (2.37). However, in this case

$$\mu = \frac{m_2}{m_1 + m_2}.$$

This particular map is not obtained from a map $\mathcal{M}$ satisfying Theorem 2.17 with an elliptic fixed point. It is derived directly from the flow for the restricted problem in action,angle variables. In section 3.4 we will show that the restricted problem does give rise to a map of the form (2.45), different than the one defined on $\sum$.

Set $\mu = 0$ and choose an invariant curve $\gamma_0(r)$ satisfying

$$\lambda(r) = \frac{n}{m},$$

$n > 0, m > 0, n, m \in \mathbb{Z}$. Each point of the circle $r =$ constant is a fixed point of the mth iterate of M_0, M_0^m. The map M_0^m is parabolic since the points of the circle S^1 are mapped into themselves, and $\lambda_1 = \lambda_2 = 1$ is verified. When μ is sufficiently small, most of these fixed points disappear and only a finite number remain for the perturbed map M_μ^m. Poincaré proved that *generically* for μ sufficiently small, M_μ^m has $2kn$ fixed points, where k is a positive integer, which lie close to S^1 [16, 89]. It can also be proven that half of these fixed points are hyperbolic and the other half are elliptic.

Thus, in each resonance map, there generically exist an equal number of elliptic and hyperbolic fixed points for M_μ^m, obtained from Poincaré's result. The hyperbolic fixed points have one-dimensional stable and unstable hyperbolic manifolds which in general will have transverse intersections. This will give rise to the complicated hyperbolic network we described in section 2.3 and infinitely many transverse homoclinic and heteroclinic points. The stable and unstable manifolds associated to the hyperbolic fixed points weave throughout the resonance zone in an extremely complicated way, surrounding the invariant curves that will surround the kn elliptic fixed points, assuming they are of the *general elliptic type*, i.e., assuming the map M_μ in a neighborhood of each of these points can be written in the form prescribed by Theorem 2.17. Each of these points then has its own family of invariant curves with smaller resonance bands. Each of these resonance bands generally repeats the dynamics described about $r = 0$. We will refer to the pattern of a family of invariant curves about an elliptic fixed point as an *elliptic island* or *tree ring pattern*. Thus, a hierarchy of ring patterns, homoclinic points, and hyperbolic points is obtained together with an intricate hyperbolic network in each resonance zone about $r = 0$. Within each resonance zone, the entire pattern repeats infinitely often. Figure 2.14 gives a sense of the complexity of the geometry [89, 204, 175, 16].

An important theorem of Zehnder [231] proves that the assumption of $\mathcal{M}$ being written in the form (2.45) is generic and that in any neighborhood of $(0,0)$ there exists a transverse homoclinic point. This generically proves the existence of the complicated dynamics of $\mathcal{M}$ just described in $\tilde{\Lambda}$ due to the hyperbolic network that results.

Theorem 2.22 (Zehnder) *Let $\mathcal{M}$ be an area-preserving map of class C^k, $k \geq 1$, or C^∞, with $(0,0)$ as an elliptic fixed point. Then any neighborhood of $\mathcal{M}$, in the space of area-preserving maps, contains an area-preserving map with the following properties, (i) $(0,0)$ is an elliptic fixed point with an eigenvalue not a root of unity; (ii) every neighborhood of $(0,0)$ contains a transverse homoclinic point.*

Property (i) implies that $\lambda_1^k \neq 1, k \geq 1$. This is more than what is assumed in Theorem 2.17, where only a finite number of conditions on λ_1

are required. Thus (2.45) can generically be achieved. A notion of distance in the space of area-preserving maps is given by Zehnder by defining a natural topology [175]. The term *generic* is used here to mean that for any given map under the given assumptions in Theorem 2.22, the assertions of the theorem are true in arbitrarily small neighborhoods of the map, within the space of area-preserving maps, and are not necessarily true for the given map itself. However, one would expect the assertions to be true in general for the given map if they are true in arbitrarily small neighborhoods.

Figure 2.14 Complicated dynamics in a neighborhood of an elliptic point.

The dynamics of $\mathcal{M}$ on the set $\tilde{\Lambda}$ is clearly very intricate. Iterates of $\mathcal{M}$ hop forever throughout the resonance zone within the infinitely complex maze.

It is remarkable that this situation is realized by the restricted problem, as is proven in the next section.

2.4 PERIODIC ORBITS AND ELLIPTIC FIXED POINTS

In section 2.1 we proved the existence of quasi-periodic motion for P_3 moving about P_1 using the KAM theorem applied to elliptic Kepler motion about P_1, where it was necessary to assume $e > 0$ in Theorem 2.3, when $\mu = 0$. That is, circular Kepler orbits were excluded. This is because Delaunay variables are not defined for $e = 0$. It is precisely this case that is of particular interest.

We now consider the case where P_3 moves about P_1 in a circular orbit for $\mu = 0$, with a given period T_0. This motion can be reduced to an area-preserving map $M(\mu)$, for μ sufficiently small, which lies near to the map M_μ for $0 < e \ll 1$. It is recalled that M_μ is the map of the form (2.37) we constructed in section 2.2 for the restricted problem on the section $\sum$ to the two-dimensional tori proven to exist in Theorem 2.3. However, $M_\mu \not\to M(\mu)$ as $e \to 0$ [204].

$M(\mu)$ is constructed on each Jacobi surface $\tilde{J}^{-1}(C)$ by first continuing the circular periodic orbits for $\mu = 0$ into a family $\mathcal{F}(\mu)$ of periodic orbits about P_1 for μ sufficiently small. The map $M(\mu)$ will be constructed from $\mathcal{F}(\mu)$ on a suitable two-dimensional surface of section.

The orbits in $\mathcal{F}(\mu)$ will also have period T_0. By Kepler's third law, T_0 is a function of the radius r_{13} (i.e. semi-major axis a) of the circular orbit for $\mu = 0$, which in turn yields the Kepler energy $\tilde{H} = -1/2a$, and the value of $\tilde{J}$ for $\mu = 0$ is given by (1.71), which we label $\tilde{J} = \tilde{J}_0 = C_0$. A given value of $a = r_{13}$ determines $C = C_0$. When each of these circular orbits are continued, for μ sufficiently small, $\tilde{J} = C(\mu) = C_0 + \mathcal{O}(\mu)$. Each orbit $\phi(t,\mu)$ in $\mathcal{F}(\mu)$ will lie on the surface $\tilde{J}^{-1}(C(\mu))$ and have period T_0. The initial values of $\phi(t,\mu)$, $\phi(0,\mu)$, will be uniquely determined as smooth functions of μ to obtain the periodic orbits of $\mathcal{F}(\mu)$. This construction is now carried out in detail.

It is necessary to restrict the circular frequencies $\tilde{\omega} = 2\pi/T_0$ so that

$$\tilde{\omega} \neq n^{-1}, \qquad n = \pm 1, \pm 2, \ldots, \tag{2.48}$$

or equivalently

$$T_0 \neq 2\pi|n|, \tag{2.49}$$

which corresponds to orbits of P_3 for $\mu = 0$ in resonance with m_2.

Let $\mathcal{F}^0(\tilde{\omega})$ denote the set of circular periodic orbits for the Kepler problem in rotating coordinates for P_3 about P_1 for $\mu = 0$, in (1.61). These orbits are identical to the circular Kepler orbits in inertial coordinates since they are invariant under rotation about the origin which was used to obtain (1.61), where $\omega = 1$. Kepler's third law implies $r_{13}^3(\tilde{\omega}+1)^2 = 1$, and to insure that P_3 for $\mu = 0$, and therefore for μ sufficiently small by continuity, does not collide with P_3, i.e., $r_{13} \neq 1$, we assume $\tilde{\omega} \neq -2, 0$. Also, to assume r_{13} is bounded, $\tilde{\omega} \neq -1$. Collectively,

$$\tilde{\omega} \neq -2, -1, 0. \tag{2.50}$$

Theorem 2.23 (Poincaré) *The family $\mathcal{F}^0(\tilde{\omega})$ of circular periodic orbits for $\mu = 0$ with Jacobi energy $\tilde{J}_0 = C_0$ and period T_0 can be uniquely continued in a smooth fashion as a function of μ for μ sufficiently small into a family*

$\mathcal{F}(\mu,\tilde{\omega})$ of periodic orbits of period T_0 lying on the Jacobi surface $\tilde{J}^{-1}(C(\mu))$, provided (2.48), (2.50) are satisfied, where $\mathcal{F}(0,\tilde{\omega}) = \mathcal{F}^0(\tilde{\omega}), C(0) = C_0$.

Proof. The method of proof was devised by Poincaré [190]. It is called the *Poincaré continuation method* and relies on the use of the implicit function theorem.

$\mathcal{F}^0(\tilde{\omega})$ is given by the orbits $\boldsymbol{\phi}^0(t) = (x_1(t), x_2(t), x_3(t), x_4(t))$, where

$$x_1 = r_{13}c, \; x_2 = r_{13}s, \; x_3 = -r_{13}\tilde{\omega}s, \; x_4 = r_{13}\tilde{\omega}c, \tag{2.51}$$

$c \equiv \cos\tilde{\omega}t, s \equiv \sin\tilde{\omega}t, \tilde{\omega} = \pm r_{13}^{-\frac{3}{2}} - 1, \dot{x}_1^2 + \dot{x}_2^2 = r_{13}^2\tilde{\omega}^2$. We write (1.61) as a first order system $\dot{\mathbf{x}} = \mathbf{f}(\mathbf{x},\mu), \mathbf{f} = (f_1, f_2, f_3, f_4) \in \mathbf{R}^4$, which in component form is

$$\begin{aligned} \dot{x}_1 &= f_1 = x_3, \\ \dot{x}_2 &= f_2 = x_4, \\ \dot{x}_3 &= f_3 = 2x_4 + x_1 + \tilde{\Omega}_{x_1}, \\ \dot{x}_4 &= f_4 = -2x_3 + x_2 + \tilde{\Omega}_{x_2}, \end{aligned} \tag{2.52}$$

where $\tilde{\Omega}$ is defined in (1.62).

$\boldsymbol{\phi}^0(t) \in \tilde{J}_0^{-1}(C_0)$ are the only orbits invariant under rotation about the origin in the x_1, x_2-plane for the Kepler problem. The only other case, in three dimensions, is the family of bounded consecutive collision orbits which lie on the x_3-axis, perpendicular to the x_1, x_2-plane, passing through the origin [23]. The value of C_0 is readily computed by substitution of (2.51) into (1.63).

A smooth continuation $\boldsymbol{\phi}(t,\mu) = \mathbf{x}(t, \mathbf{x}(0,\mu), \mu)$ of periodic orbits with respect to μ of $\boldsymbol{\phi}$ is sought on the surface $\tilde{J}^{-1}(C(\mu))$, for μ sufficiently small, where $\boldsymbol{\phi}(t,\mu)$ have period T_0 and $\boldsymbol{\phi}(t,0) = \mathbf{x}(t,\mathbf{x}(0,0),0) \equiv \boldsymbol{\phi}^0(t), \mathbf{x}(0,0) = \boldsymbol{\phi}^0(0)$. $\mathbf{x}(0,\mu), C(\mu)$ can be written as a convergent power series in μ about $\mu = 0$.

We set

$$\mathbf{F}(t, \mathbf{x}(0,\mu), \mu) \equiv \mathbf{x}(t, \mathbf{x}(0,\mu), \mu) - \mathbf{x}(0,\mu).$$

Thus, we would like to solve the system of equations

$$F_k(T_0, \mathbf{x}(0,\mu), \mu) = x_k(T_0, \mathbf{x}(0,\mu), \mu) - x_k(0,\mu) = 0, \tag{2.53}$$

$k = 1,2,3,4$. Equation (2.53) is satisfied for $\mu = 0, \mathbf{x}(0,0) = \boldsymbol{\phi}^0(0)$, since

$$F_k(T_0, \mathbf{x}(0,0), 0) = \phi_k^0(T_0) - \phi_k^0(0) = 0.$$

A unique solution $\mathbf{x}(0,\mu)$ of (2.53) can be found for μ sufficiently small if

$$\det\left(\frac{\partial F_k(t, \mathbf{x}(0,\mu), \mu)}{\partial x_\ell(0,\mu)}\right)\Bigg|_{\substack{t = T_0 \\ \mu = 0 \\ \mathbf{x}(0,0) = \boldsymbol{\phi}^0(0)}} \neq 0\,, \tag{2.54}$$

$k, \ell = 1, 2, 3, 4$, which follows by the implicit function theorem and by continuity with respect to μ of the flow for μ sufficiently small. For notation, $\partial a_i/\partial b_j, i = 1, 2, \ldots, n, j = 1, 2, \ldots, m$, represents the $n \times m$ matrix $\mathbf{A}$ with elements $\alpha_{ij} = \partial a_i/\partial b_j$, which we write as $\mathbf{A} = (\alpha_{ij})$.

We set

$$\mathbf{B} = \left(\frac{\partial F_k(t, \mathbf{x}(0,\mu), \mu)}{\partial x_\ell(0,\mu)} \right) \Bigg|_{\substack{t = T_0 \\ \mu = 0 \\ \mathbf{x}(0,0) = \boldsymbol{\phi}^0(0)}} \tag{2.55}$$

$$\mathbf{C} = \left(\frac{\partial x_k(t, \mathbf{x}(0,\mu), \mu)}{\partial x_\ell(0,\mu)} \right) \Bigg|_{\substack{t = T_0 \\ \mu = 0 \\ \mathbf{x}(0,0) = \boldsymbol{\phi}^0(0)}} . \tag{2.56}$$

Then by the definition of $\mathbf{F}$,

$$\mathbf{B} = \mathbf{C} - \mathbf{I}_4, \tag{2.57}$$

$\mathbf{I}_4$ being the 4×4 identity matrix. For notation, we set $\det(\mathbf{A}) = |\mathbf{A}|$ for any square matrix $\mathbf{A}$.

Equation (2.54) is not satisfied. That is,

$$|\mathbf{B}| = |\mathbf{C} - \mathbf{I}_4| = 0. \tag{2.58}$$

Thus, $\mathbf{B}$ is singular.

We can prove $\mathbf{B}$ is singular in two different ways. One is that $\boldsymbol{\phi}^0(t)$ is a periodic solution for (2.52) for $\mu = 0$ together with the fact that (2.52) is autonomous. The other is because $\boldsymbol{\phi}^0(t)$ is periodic and $\tilde{J}$ is an integral. Both of these lead to $|\mathbf{B}| = 0$. Understanding how this occurs will show how to modify $\mathbf{B}$ to obtain a nonsingular matrix. This is now examined.

Consider the restricted problem (2.52). Let $\boldsymbol{\zeta}(t)$ be a periodic solution for a given value of $\mu = \mu^*$. For example, for $\mu^* = 0$, we consider $\boldsymbol{\zeta}(t) = \boldsymbol{\phi}^0(t)$. For a general solution $\mathbf{x}(t, \mathbf{x}(0), \mu)$ of (2.52),

$$\tilde{J}(\mathbf{x}(t, \mathbf{x}(0), \mu), \mu) = \tilde{J}(\mathbf{x}(0), \mu).$$

Differentiating this with respect to $x_\ell(0)$ implies

$$(\tilde{J}_{\mathbf{x}}, \mathbf{x}_{x_\ell(0)}) = \tilde{J}_{x_\ell(0)}.$$

Setting $\mu = 0, \mathbf{x}(0) = \boldsymbol{\phi}^0(0), t = T_0$ implies

$$\tilde{J}_{\mathbf{x}}(\boldsymbol{\phi}^0(0), 0)\mathbf{C} - \tilde{J}_{\mathbf{x}}(\boldsymbol{\phi}^0(0), 0) = 0. \tag{2.59}$$

Assuming $\tilde{J}_{\mathbf{x}}(\boldsymbol{\phi}^{\mathbf{0}}(0),0) \neq \mathbf{0}$, which is indeed satisfied, then (2.59) implies

$$|\mathbf{C} - \mathbf{I}_4| = 0, \tag{2.60}$$

yielding (2.58). In general, (2.59) implies that 1 is a left eigenvalue of $\mathbf{C}$ with left eigenvector $\tilde{J}_{\mathbf{x}}(\boldsymbol{\phi}(0),0)$, which means that $\tilde{J}_{\mathbf{x}}(\boldsymbol{\phi}(0),0)$ is mapped into itself by right multiplication with $\mathbf{C}$.

An eigenvalue of 1 for $\mathbf{C}$ is indicative of the existence of an invariant quantity under the flow of (2.52), in this case $\tilde{J}$.

Now, set

$$\mathbf{X}(t) = \left(\frac{\partial x_\ell}{\partial x_j(0)}\right), \tag{2.61}$$

$j, \ell = 1,2,3,4$, where $\mathbf{x}(t,\mathbf{x}(0),\mu))$ is a general solution of (2.52), $\mathbf{X}(0) = \mathbf{I}_4$. We set $\mu = 0$ and $\mathbf{x}(0) = \boldsymbol{\phi}^{\mathbf{0}}(0)$, so that $\mathbf{x}(t,\mathbf{x}(0),0) = \boldsymbol{\phi}^{\mathbf{0}}(t,\boldsymbol{\phi}^{\mathbf{0}}(0))$. It is verified that the columns $\mathbf{X}_i(t)$ of the matrix $\mathbf{X} = (\mathbf{X}_1,\mathbf{X}_2,\mathbf{X}_3,\mathbf{X}_4)$ satisfy the first order system

$$\dot{\mathbf{X}} = \mathbf{A}(t)\mathbf{X}, \quad \mathbf{A} = (\partial f_i(\boldsymbol{\phi}^{\mathbf{0}}(t),0)/\partial x_j), \tag{2.62}$$

which is called the *variational equation* for the variation of solutions $\mathbf{x}(t)$ of (2.52) for $\mu = 0$ about $\boldsymbol{\phi}^{\mathbf{0}}(t)$. $\mathbf{A}$ is a 4×4 matrix which is periodic of period T_0. Let $\mathbf{X}_i(t)$ be solutions of (2.62) satisfying $\mathbf{X}_i(0) = \hat{\mathbf{e}}_i$, where $\hat{\mathbf{e}}_1 = (1,0,0,0), \hat{\mathbf{e}}_2 = (0,1,0,0)$, etc. Then $\mathbf{X}(t)$ is called a *fundamental solution matrix*, which we label as $\boldsymbol{\Phi}(t)$.

The matrix $\boldsymbol{\Phi}(t)$, satisfying $\boldsymbol{\Phi}(0) = I_4$, is given by

$$\boldsymbol{\Phi}(t) = \left(\frac{\partial x_k(t)}{\partial x_\ell(0)}\right)\Big|_{\mathbf{x}(0)=\boldsymbol{\phi}^{\mathbf{0}}(0)} \tag{2.63}$$

and is a matrix solution to (2.62), since each of its columns are a solution of (2.62).

$\boldsymbol{\Phi}(t)$ is called the *first variation* of $\mathbf{x}(t) \equiv \mathbf{x}(t,\mathbf{x}(0),0)$ about $\boldsymbol{\phi}^{\mathbf{0}}(t,\boldsymbol{\phi}(0))$. This is seen by expanding $\mathbf{x}(t)$ about $\boldsymbol{\phi}^{\mathbf{0}}(t)$; then

$$\mathbf{x}(t) = \boldsymbol{\phi}^{\mathbf{0}}(t) + \boldsymbol{\Phi}(t)(\mathbf{x}(0) - \boldsymbol{\phi}^{\mathbf{0}}(0)) + \mathcal{O}(|\mathbf{x}(0) - \boldsymbol{\phi}^{\mathbf{0}}(0)|^2), \tag{2.64}$$

as is verified. It is seen that

$$\mathbf{C} = \boldsymbol{\Phi}(T_0). \tag{2.65}$$

Thus, $\mathbf{C}$ can be computed when $\boldsymbol{\Phi}(t)$ can be obtained. The computation of the form of general solutions of linear systems of differential equations with periodic coefficients is called *Floquet theory* [116, 60]. The general form of the solutions can be explicitly determined, and they take on particularly simple form when $\mathbf{C}$ has a complete eigenspace.

In the four-dimensional phase space for (2.52) with coordinates $x_k, k = 1,2,3,4$, a four-dimensional element as it flows over time t with the solutions of (2.52) does not change volume, since (2.52) is conservative, which follows by Liouville's theorem. In time t, each point $\boldsymbol{\alpha}$ of the element moves to the location $\mathbf{x}(t, \boldsymbol{\alpha})$. Thus, the flow of (2.52) acting on the element yields a mapping of the element over time t. The Jacobian matrix of the map is given by $\mathbf{X}(t, \boldsymbol{\alpha})$. Since the map is volume-preserving (or *symplectic*), then $|\mathbf{X}(t)| = 1$. Thus, setting $t = T_0$ implies $|\mathbf{C}| = 1$. This can also be derived by showing $\frac{d}{dt}|\mathbf{X}(t)| = |\mathbf{X}(t)|(\text{trace}A(t))$, implying $\ln|\mathbf{X}(t)| = \int_0^t (\text{trace}A(t))dt = 0$ or $|\mathbf{X}(t)| = 1$. See [204].

$\mathbf{C}$ is a linear symplectic map. This implies that the eigenvalues of $\mathbf{C}$ must occur in reciprocal pairs. That is, if λ is an eigenvalue, then λ^{-1} is an eigenvalue [15]. Thus, $\mathbf{C}$ has another eigenvalue of 1. This additional eigenvalue of 1 is due to another invariance of the flow of (2.52) and follows from the following lemma.

Lemma 2.24

$$\mathbf{Cf} = \mathbf{f}, \tag{2.66}$$

where $\mathbf{Cf} \equiv \sum_{j=1}^{4} C_{ij} f_j, f_j = f_j(\boldsymbol{\phi}^{\mathbf{0}}(0)) \neq 0.$

This is proven in [204]. It results from the fact that (2.52) is autonomous. This implies that if $\boldsymbol{\phi}^{\mathbf{0}}(t, \boldsymbol{\phi}^{\mathbf{0}}(0))$ is a periodic orbit for $\mu = 0$, then $\boldsymbol{\psi}(t, \boldsymbol{\psi}(0)) = \boldsymbol{\phi}^{\mathbf{0}}(t + \tau, \boldsymbol{\psi}(0)), \boldsymbol{\psi}(0) = \boldsymbol{\phi}^{\mathbf{0}}(\tau, \boldsymbol{\phi}^{\mathbf{0}}(0)), \tau > 0$, is also a periodic orbit, obtained by translating the initial value $\boldsymbol{\psi}(0)$ along the orbit $\boldsymbol{\phi}^{\mathbf{0}}(t, \boldsymbol{\phi}^{\mathbf{0}}(0))$.

We return to (2.53) and use the knowledge of (2.59), (2.66) to modify (2.53) so that a unique solution $\mathbf{x}(0, \mu)$ can be found for μ sufficiently small.

First, in view of the discussion of why Lemma 2.24 is true, we need to prevent the initial values $\boldsymbol{\phi}(0)$ of $\boldsymbol{\phi}^{\mathbf{0}}(t)$ from varying along the curve and restrict them to a surface of section to $\boldsymbol{\phi}^{\mathbf{0}}(t)$.

To find a section of codimension 1, we notice that from (2.51),

$$\boldsymbol{\phi}^{\mathbf{0}}(0) = (r_{13}, 0, 0, r_{13}\tilde{\omega}). \tag{2.67}$$

It is verified from (2.52) that $f_3(\boldsymbol{\phi}^{\mathbf{0}}(0), 0) = -r_{13}^2\tilde{\omega} \neq 0$. Thus $\dot{x}_3(0, \boldsymbol{\phi}^{\mathbf{0}}(0), 0) \neq 0$, and so the hyperplane $\{x_3 = \phi_3^0(0) = 0\}$ is transversal to $\boldsymbol{\phi}^{\mathbf{0}}(t)$ and therefore to the flow of (2.52) for μ sufficiently small, with initial values $\mathbf{x}(0, \mu)$ sufficiently near $\boldsymbol{\phi}^{\mathbf{0}}(0)$. We only allow $\mathbf{x}(0, \mu)$ to vary on the three-dimensional plane $\{x_3 = 0\}$. That is, only $x_1(0, \mu), x_2(0, \mu), x_4(0, \mu)$ are allowed to vary.

Let S denote the plane defined by $\{x_3 = 0\}$. Now, with $x_3 = \phi_3^0(0) = 0$ on S, for $t = 0$, this means the variable $x_3(0, \mu)$ is deleted in (2.53), which now only has the three unknowns $x_k(0,\mu), k = 1,2,4$ on S. This implies that the *third column* of $\mathbf{B}$ is deleted.

We now need three equations to solve, so one needs to be deleted to obtain a unique solution. It turns out that $F_4(T_0, \mathbf{x}(0,\mu), \mu) = 0$ can be omitted because the Jacobi integral can be used to show it is indeed satisfied. It is verified that $\tilde{J}_{x_4}(\phi^0(0), 0) = r_{13}\tilde{\omega} \neq 0$. $F_4 = 0$ follows by the *mean value theorem*, since

$$\begin{aligned}\hat{J} &\equiv \tilde{J}(\mathbf{x}(T_0, \mathbf{x}(0,\mu),\mu),\mu) - \tilde{J}(\mathbf{x}(0,\mu),\mu)\\ &= \tilde{J}_{x_4}(\gamma,\mu)F_4(T_0,\mathbf{x}(0,\mu),\mu),\end{aligned}$$

where $x_4(0,\mu) < \gamma_4 < x_4(T_0, \mathbf{x}(0,\mu),\mu)$, and $\gamma_j = x_j(0,\mu), j = 1,2,3, \gamma = (\gamma_1, \ldots, \gamma_4)$. On the other hand, $\hat{J} = 0$, $\tilde{J}_{x_4}(\gamma,\mu) \neq 0$ and thus $F_4(T_0, \mathbf{x}(0,\mu), \mu) = 0$. This implies that the *fourth row* of $\mathbf{B}$ can be omitted.

Thus, *instead of satisfying (2.53)*, we need only satisfy (2.53) with the fourth equation omitted and the independent variable $x_3(0,\mu) = 0$. This yields three equations for $F_k = 0, k = 1,2,3$ for the three variables $x_j(0,\mu)$, $j = 1,2,4$.

Hence instead of trying to satisfy $|\mathbf{B}| \neq 0$, which is not true, we only need to consider the reduced 3×3 matrix $\mathbf{D}$ obtained by *deleting the third column and fourth row from* $\mathbf{B}$.

A lengthy calculation verifies that

$$|\mathbf{D}| = 24\pi \sin^2(\pi\tilde{\omega}^{-1}), \tag{2.68}$$

which is nonzero provided (2.48) is satisfied [204]. This proves Theorem 2.23. □

2.4.1 Solution of the Variational Equations

The proof of Theorem 2.23 follows the outline in [204], where the calculation of $\mathbf{D}$ is not shown. In order to calculate $\mathbf{D}$, the variational equations (2.62) need to be solved. The solution of (2.62) is a lengthy process and without it the proof cannot be completed.

For completeness, the solution of (2.62) is outlined.

We consider the restricted problem (2.52) for $\mu = 0$, which represents the Kepler problem in rotating coordinates. We require the variational system (2.62) for this problem.

It is verified that in the coordinates $x_k, k = 1,2,3,4$, $\mathbf{A}(t)$ depends explicitly on time. Such a system is not easy to solve, and Floquet theory is required. The key is to obtain a variational system where $\mathbf{A}(t)$ reduces to a constant matrix. There are several ways to achieve this.

One method is to use polar coordinates, $x_1 = r\cos\theta, x_2 = r\sin\theta$. Writing (2.52) as a second-order system of differential equations in the variables x_1, x_2, one obtains

$$\ddot{r} - r(\dot{\theta}+1)^2 = -r^{-2}, \quad 2\dot{r}(\dot{\theta}+1) + r\ddot{\theta} = 0. \tag{2.69}$$

In these variables the time dependence in $\phi^0(t)$ nearly disappears, and the solution becomes $\tilde{\phi}^0(t) = (r_{13}, \tilde{\omega}t, 0, \tilde{\omega})$. Setting $x_1 = r, x_2 = \theta, x_3 = \dot{r}, x_4 = \dot{\theta}$ and writing (2.69) as a first order system, it is verified that the variational system $\dot{\mathbf{X}} = \mathbf{A}\mathbf{X}$ is obtained where $\mathbf{A}$ is a constant matrix

$$\mathbf{A} = \begin{pmatrix} \mathbf{O}_2 & \mathbf{I}_2 \\ \mathbf{A}_1 & \mathbf{A}_2 \end{pmatrix},$$

$\mathbf{O}_2$ is the 2×2 zero matrix, $\mathbf{I}_2$ is the 2×2 identity matrix,

$$\mathbf{A}_1 = \begin{pmatrix} 3(\tilde{\omega}+1)^2 & 0 \\ 0 & 0 \end{pmatrix},$$

$$\mathbf{A}_2 = \begin{pmatrix} 0 & 2r_{13}(\tilde{\omega}+1) \\ -2r_{13}^{-1}(\tilde{\omega}+1) & 0 \end{pmatrix}.$$

Since $\mathbf{A}$ is a constant matrix, the variational system can be explicitly solved by standard methods once the eigenvalue and eigenvectors are computed. The general solution $X(t) \in \mathbf{R}^4$ of the variational system is represented as a linear combination of the eigenvectors with unknown constants as coefficients. The constants are successively determined for each $i = 1, 2, 3, 4$ by setting $\mathbf{X}(0) = \hat{\mathbf{e}}_i$. For each i they are uniquely determined, which yields the respective solution $\mathbf{X}_i(t)$. The $\mathbf{X}_i(t)$ form the columns of the fundamental solution matrix.

The use of polar coordinates to compute the solution of the variational equation has a drawback since the variational matrix we seek is in Cartesian coordinates. The transformation to Cartesian coordinates of the variational matrix isn't convenient.

There is another choice of coordinates that also reduces $\mathbf{A}$ to a constant form, used in [204]. These coordinates rotate with the same uniform rate as the solution $\phi^0(t)$, and thus, in them the periodic solution is a constant. Although they are not quite as simple as the polar coordinates, it is straight forward to obtain the variational matrix in Cartesian coordinates from them; we will also use them to be consistent with the proof of Theorem 2.23, which conforms to [204]. The solution of the variational equations is not carried out in [204], and this is done here.

The change of coordinates is given by

$$\mathbf{y} = \tilde{\mathbf{R}}(\tilde{\omega}t)\mathbf{x} \tag{2.70}$$

with inverse

$$\mathbf{x} = \tilde{\mathbf{R}}^{-1}(\tilde{\omega}t)\mathbf{y},$$
$$\tilde{\mathbf{R}} = \begin{pmatrix} \mathbf{R} & \mathbf{O}_2 \\ \mathbf{O}_2 & \mathbf{R} \end{pmatrix}, \tag{2.71}$$
$$\tilde{\mathbf{R}}^{-1} = \begin{pmatrix} \mathbf{R}^{-1} & \mathbf{O}_2 \\ \mathbf{O}_2 & \mathbf{R}^{-1} \end{pmatrix};$$

$\mathbf{R} = \mathbf{R}(\tilde{\omega}t), \mathbf{R}^{-1} = \mathbf{R}^{-1}(\tilde{\omega}t)$ are given by

$$\mathbf{R} = \begin{pmatrix} c & s \\ -s & c \end{pmatrix}, \quad \mathbf{R}^{-1} = \begin{pmatrix} c & -s \\ s & c \end{pmatrix}, \tag{2.72}$$

$c = \cos\tilde{\omega}t, s = \sin\tilde{\omega}t$. Equation (2.52) is transformed into

$$\dot{\mathbf{y}} = \begin{pmatrix} \tilde{\omega}\mathbf{J} & \mathbf{I}_2 \\ (1-b)\mathbf{I}_2 & (2+\tilde{\omega})\mathbf{J} \end{pmatrix}\mathbf{y}, \tag{2.73}$$

where $b = r_{13}^{-3} = (\tilde{\omega}+1)^2$,

$$\mathbf{J} = \begin{pmatrix} 0 & 1 \\ -1 & 0 \end{pmatrix}.$$

$\phi^{\mathbf{0}}(t)$ is transformed into

$$\tilde{\phi}^{\mathbf{0}}(t) = (r_{13}, 0, 0, r_{13}\tilde{\omega}), \tag{2.74}$$

which is a *constant vector.* The variational system relative to $\tilde{\phi}(t)$ is verified to be

$$\dot{\mathbf{Y}} = \tilde{\mathbf{A}}\mathbf{Y}, \tag{2.75}$$

$\mathbf{Y}(0) = \mathbf{I}_4 \equiv 4 \times 4$ identity matrix, where

$$\tilde{\mathbf{A}} = \begin{pmatrix} \tilde{\omega}\tilde{J} & \mathbf{I}_2 \\ \mathbf{B} & (2+\tilde{\omega})\mathbf{J} \end{pmatrix},$$
$$\mathbf{B} = \begin{pmatrix} 1+2b & 0 \\ 0 & 1-b \end{pmatrix}.$$

$\tilde{\mathbf{A}}$ is a *constant* 4×4 *matrix.* The general matrix solution $\mathbf{Y}(t)$ to (2.75) can therefore be found by standard methods using an eigenvector decomposition [16]. The eigenvalues $\lambda = 0, 0, \pm i(\tilde{\omega}+1)$ are first obtained for $\tilde{\mathbf{A}}$, where $\lambda = 0$ has only a one-dimensional eigenspace. The eigenvector $\mathbf{v}_1$ corresponding to $\lambda_1 = \lambda_2 = 0$ is $\mathbf{v}_1 = (0, -\tilde{\omega}^{-1}, 1, 0)$. The eigenvectors $\mathbf{v}_3, \mathbf{v}_4$ corresponding to $\lambda_3 = i(\tilde{\omega}+1), \lambda_4 = -i(\tilde{\omega}+1)$ are $\mathbf{v}_3 = (1-2b, 2ib(1-b)^{-1}, i(\tilde{\omega}+1), \tilde{\omega}), \mathbf{v}_4 = \bar{\mathbf{v}}_3$, after some calculations and simplifications. The general form of the real vector solution $\varphi(t) \in \mathbf{R}^4$ for (2.75) can be written in the form [16]

$$\varphi(t) = c_1\mathbf{v}_1 + c_2(\mathbf{v}_2 + \mathbf{v}_1 t) + e^{i(\tilde{\omega}+1)t}[f\mathbf{v}_3 + \bar{g}\bar{\mathbf{v}}_u]$$
$$+e^{-i(\tilde{\omega}+1)t}[f\mathbf{v}_4 + \bar{g}\bar{\mathbf{v}}_3],$$

where c_1, c_2 are real constants, f, g are complex constants, $f = f_1 + if_2, g = g_1 + ig_2$, and $\mathbf{A}\mathbf{v}_2 = \mathbf{v}_1$.

The columns $\mathbf{Y}_i(t)$ of the fundamental solution matrix $\tilde{\mathbf{\Phi}}(t)$ of (2.75) are obtained by solving uniquely for c_1, c_2, f, g when

$$\boldsymbol{\varphi}(0) = \hat{\mathbf{e}}_i,$$

for each $i = 1, 2, 3, 4$, and then for each i, $\mathbf{Y}_i(t) = \boldsymbol{\varphi}_i(t)$. The fundamental solution matrix $\mathbf{\Phi}(t)$ for (2.62) is verified to be

$$\mathbf{\Phi}(t) = \tilde{\mathbf{R}}^{-1}(\tilde{\omega}t)\tilde{\mathbf{\Phi}}(t),$$

and $\mathbf{C}$ is obtained from (2.65), setting $t = T_0$. Thus, $\mathbf{\Phi}(T_0) = \tilde{\mathbf{\Phi}}(T_0)$, since $\tilde{\mathbf{R}}^{-1}(\tilde{\omega}T_0) = \mathbf{I}_4$. The calculation of (2.68) directly follows.

Theorem 2.23 yields periodic solutions with the same period T_0 as $\boldsymbol{\phi}^0(t)$. Other families of solutions can be obtained from $\mathcal{F}(\mu, T_0)$. For example, let $\boldsymbol{\phi}(t, \tilde{\mu}) \in \mathcal{F}(\tilde{\mu}, T_0)$, having corresponding constant $\tilde{C} = C(\tilde{\mu})$, where $\mu = \tilde{\mu}$ is now fixed sufficiently small. It can be proven that for each value of $J = C$, where C is sufficiently near $\tilde{C}$, there exists a periodic solution of period $T(C)$ near to $T_0, T(\tilde{C}) = T_0$ [204]. This yields a new family of periodic orbits $\tilde{\mathcal{F}}(C)$ for $\mu = \tilde{\mu}$ with period $T(C)$. Thus, in the family $\tilde{\mathcal{F}}(C)$, the period is not fixed to T_0, and, moreover, can be expanded as a convergent power series in C for C sufficiently near $\tilde{C}$. $\tilde{\mathcal{F}}(C)$ bifurcates from $\mathcal{F}(\mu)$ at $\mu = \tilde{\mu}$.

The periodic orbits of the restricted problem given by $\mathcal{F}(\mu)$ which are continuations at the circular periodic orbits for the Kepler problem, for μ sufficiently small, play a prominent role in the work of Poincaré in celestial mechanics. He labeled them *periodic orbits of the first kind.* As remarked earlier, analogous to $\boldsymbol{\phi}^0(t)$ in the plane the consecutive collision orbits lying on the x_3-axis for the three-dimensional Kepler problem are also invariant under rotation about the origin in the x_1, x_2-plane. They can also be proven to give rise to a smooth continuation of a family of periodic orbits for the three-dimensional restricted problem for μ sufficiently small [23, 158]. In this case the collision orbit needs to be regularized, and the geodesic flow equivalence of the Kepler problem in section 1.6 is utilized. This family of periodic orbits therefore can be viewed as another type of periodic orbit of the first kind.

2.4.2 An Area-Preserving Map near an Elliptic Fixed Point

The periodic orbits in $\mathcal{F}(\mu)$ can be used to construct a special two-dimensional area-preserving map with interesting properties. Let $\boldsymbol{\phi}(t, \mu) \in \mathcal{F}(\mu)$. $\boldsymbol{\phi}(t, \mu)$ lies near $\boldsymbol{\phi}^0(t)$ for μ sufficiently small, for $0 \le t \le T_0 = 2\pi/\tilde{\omega}$, and moreover for all $t \in \mathbf{R}^1$. A two-dimensional surface of section $\sum$ to $\boldsymbol{\phi}(t, \mu)$ for $t = 0$ can be constructed with coordinates $x_1, x_3 = \dot{x}_1$, where $\sum$ is defined by $x_2 = 0$ and where $\dot{x}_2$ is determined from the Jacobi integral as a function of x_1, x_3 on $\sum$. To show that $x_4 = \dot{x}_2$ can be eliminated on $\sum$ by using $\tilde{J}(\mathbf{x}, \mu)$ defined by (1.63), it is verified that for $\mu = 0, t = 0$,

$$\tilde{J}_{x_4}(\boldsymbol{\phi}^0(0), 0) = -2\phi_4^0(0) = -2r_{13}\tilde{\omega} \neq 0. \tag{2.76}$$

Thus, by continuity with respect to μ, this is true for μ sufficiently small where $\boldsymbol{\phi}^0(0)$ is replaced by $\boldsymbol{\phi}(0, \mu)$. Thus, in a neighborhood of $\boldsymbol{\phi}(0, \mu)$ on $\sum$, x_4 can be solved as a function f of x_1, x_3, and $x_2 = 0$ by the implicit function theorem from the relationship $\tilde{J}(\mathbf{x}, \mu) = C$.

By the smoothness of solutions of (2.52) with respect to initial conditions, a point $x_1(0), x_3(0)$ on $\sum$ near $\boldsymbol{\phi}(0, \mu)$ is mapped over a time T near $T_0, T = T_0 + \mathcal{O}(\mu)$ into another point $x_1(T), x_3(T)$. This defines a map $M(\mu)$: $\sum \to \sum$ near $(\phi_1(0, \mu), \phi_3(0, \mu))$. (See Figure 2.15.) The coordinate change $y_k = x_k - \phi_k, k = 1, 2, 3, 4$, enables $M(\mu)$ to be written in the $\mathbf{y}$ coordinates with $\mathbf{y} = \mathbf{0}$ being a fixed point at the origin corresponding to $\boldsymbol{\phi}(0, \mu)$.

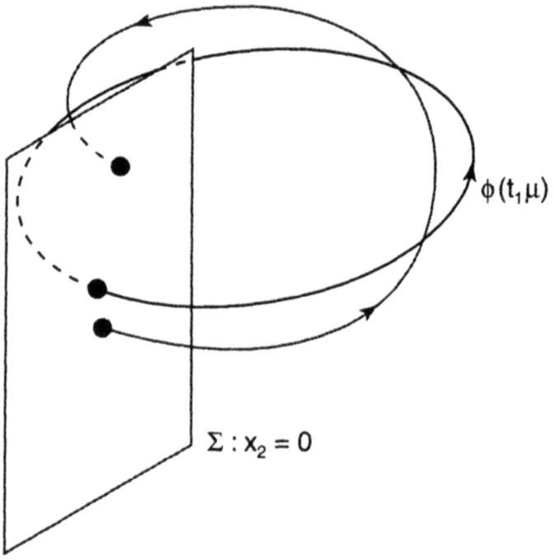

Figure 2.15 Surface of section.

From (2.64) we can obtain an expression for $M(0)$. Equation (2.64) implies

$$\mathbf{y}(T) = \mathbf{\Phi}(T_0)\mathbf{y}(0) + \mathcal{O}(|\mathbf{y}(0)|^2) + \mathcal{O}(|T - T_0||\mathbf{y}(0)|). \tag{2.77}$$

$M(0)$, to first order, is explicitly determined. Since

$$\mathbf{y}(T_0) = \mathbf{C}\mathbf{y}(0) + \cdots, \tag{2.78}$$

where we know $\mathbf{C}$ explicitly, this yields the map $M(0)$:

$$\begin{aligned} \xi_1 &= c\xi + (\tilde{\omega} + 1)^{-1} s\eta + \cdots, \\ \eta_1 &= -(\tilde{\omega} + 1)s\xi + c\eta + \cdots, \end{aligned} \tag{2.79}$$

as is verified, or

$$\zeta_1 = \mathbf{A}\zeta + \cdots, \tag{2.80}$$

where $\boldsymbol{\zeta}_1 = (\xi_1, \eta_1), \boldsymbol{\zeta} = (\xi, \eta), \xi \equiv y_1(0), \eta \equiv y_3(0), \xi_1 \equiv y_1(T), \eta_1 \equiv y_3(T)$,

$$\mathbf{A} = \begin{pmatrix} c & (\tilde{\omega}+1)^{-1}s \\ -(\tilde{\omega}+1)s & c \end{pmatrix}, \tag{2.81}$$

$c = \cos\alpha, s = \sin\alpha, \alpha = 2\pi(\frac{\tilde{\omega}+1}{\tilde{\omega}})$. It is remarked that on $\sum$, $\dot{x}_1$ represents the way a solution crosses the x_1-axis. If $\dot{x}_1 = 0$, then the orbit crosses perpendicularly, and a periodic orbit is obtained due to symmetry of solutions of the restricted problem with respect to the x_1-axis [219]. The symmetry map is $x_1 \to x_1,\ x_2 \to -x_2,\ t \to -t$. For $\dot{x}_1 \neq 0$, the solution will cross the x_1-axis in a nonperpendicular fashion, and $|\dot{x}_1|$ is a measure of that.

Equation (2.80) can also be obtained in a slightly different way using differentials and polar coordinates from (2.69) [97]. Hénon's method is actually computationally easier, and less computational since only $r(t), \theta(t)$ are solved to first order, and these solutions are then used to determine ζ_1 using differentials. Equation (2.80) is also computed in [204], where canonical variables are used, resulting in $\alpha = 2\pi/\tilde{\omega}$ being the only change. There is no advantage to using these variables. However, since higher order terms of (2.80) are computed in [204] we will also use the same canonical coordinates.

We transform $x_k, k = 1,2,3,4$, to z_1, z_2, w_1, w_2 using (2.9), yielding the Hamiltonian $\tilde{H}(\mathbf{z}, \mathbf{w}, \mu)$ and differential equations (2.10). The same notation is used for $\boldsymbol{\phi}^0(t), \boldsymbol{\phi}(t,\mu)$ in these coordinates, and $\boldsymbol{\phi}^0(0) = (r_{13}, 0, 0, r_{13}(1+\tilde{\omega}))$. Analogous to (2.76), we have $\tilde{H}_{w_2}(\boldsymbol{\phi}^0(0), 0) = r_{13}\tilde{\omega} \neq 0$ which implies $\dot{z}_2(0) \neq 0$, and $z_2 = \phi_2^0(0) = 0$ is a section $\sum$ to $\phi(t,\mu)$ at $t = 0$, with coordinates z_1, w_1 where w_2 is eliminated on $\sum$ from the relationship $\tilde{H} =$ constant corresponding to the Jacobi surface $J^{-1}(C)$ in the new coordinates. Setting $z_1(0) \equiv \xi, w_1(0) \equiv \eta, z_1(T_0) \equiv \xi_1, w_1(T_0) \equiv \eta_1$, and (2.79), (2.80) are again obtained, with $\alpha = 2\pi/\tilde{\omega}$, which we now assume.

It is remarked that (2.80) to first order is equivalent to

$$\zeta_1 = \mathbf{R}(\mathbf{T_0}) \begin{pmatrix} 1 & (\tilde{\omega}+1)^{-1} \\ -(\tilde{\omega}+1) & 1 \end{pmatrix} \zeta.$$

It represents the map $M(0)$ near the circular solution $\phi^0(t)$ for the Kepler problem in canonical rotating coordinates z_1, z_2, w_1, w_2. To see why it has this form, the general solution $\mathbf{W}(t) = (z_1(t), z_2(t), w_1(t), w_2(t))$ to the Kepler problem in canonical rotating coordinates is found applying the rotation

$$\tilde{\mathbf{R}}(t) = \begin{pmatrix} \mathbf{R} & \mathbf{O} \\ \mathbf{O} & \mathbf{R} \end{pmatrix}, \quad \mathbf{R} = \begin{pmatrix} \cos t & -sint \\ -sint & \cos t \end{pmatrix}$$

to the solution $\mathbf{Q}(t) = (q_1(t), q_2(t), p_1(t), p_2(t))$ of the Kepler problem in canonical inertial coordinates (see (1.21)). That is,

$$\mathbf{W}(t) = \tilde{\mathbf{R}}(t)\mathbf{Q}(t),$$

which represents a canonical transformation, as is verified. Since $\mathbf{Q}(0) = \mathbf{W}(0)$, then

$$\mathbf{\Phi}(t) = \left(\frac{\partial W_k(t)}{\partial W_\ell(0)}\right) = \tilde{\mathbf{R}}(t)\left(\frac{\partial Q_k(t)}{\partial Q_\ell(0)}\right).$$

Setting $t = T_0, \mathbf{W}(0) = \mathbf{Q}(0) = \boldsymbol{\phi}^{\mathbf{0}}(0)$ yields

$$\mathbf{C} = \mathbf{\Phi}(T_0) = \tilde{\mathbf{R}}(T_0)\left(\frac{\partial Q_k(T_0)}{\partial Q_\ell(0)}\right).$$

Substituting this into (2.78) yields the map $M(0)$.

Now, it is immediately seen that $|\mathbf{A}| = 1$, reflecting the fact that $M(0)$ is area preserving. This map is of the elliptic type at $\zeta = 0$ since the eigenvalues λ of $\mathbf{A}$ are computed to be

$$\lambda_1 = e^{2\pi\tilde{\omega}^{-1}i}, \qquad \lambda_2 = \bar{\lambda}_1.$$

The frequency $\tilde{\omega}$ satisfies (2.48), (2.50).

In addition to (2.48), (2.50), we also assume

$$\tilde{\omega} \neq 3n^{-1}, 4n^{-1}, \qquad n = \pm 1, \pm 2, \ldots. \tag{2.82}$$

Equation (2.82) includes (2.48), (2.50) for $\tilde{\omega} \neq 0$. This is assumed since this implies

$$\lambda_1^j \neq 1, \quad j = 1, 2, 3, 4. \tag{2.83}$$

Equation (2.83) is the basic assumption for Theorem 2.17 for the case $k = 1$ assuming $\gamma_1 \neq 0$. This theorem then implies that (2.79), with $\alpha = 2\pi/\tilde{\omega}$, can be written as

$$\begin{aligned}\xi_1 &= \xi \cos w_1 - \eta \sin w_1 + \mathcal{O}_4,\\ \eta_1 &= \xi \sin w_1 + \eta \cos w_1 + \mathcal{O}_4,\end{aligned} \tag{2.84}$$

where $\mathcal{O}_4$ are power series in ξ, η containing terms of order ≥ 4 only, and

$$w_1 = \gamma_0 + \gamma_1(\xi^2 + \eta^2), \tag{2.85}$$

where $\gamma_1 \neq 0$.

It turns out that for $M(0)$ one can explicitly calculate that $\gamma_1 = -3\pi r_{13}^{-1}\tilde{\omega}^{-2} \neq 0$ [204].

2.4.3 What Happens for $\mu > 0$ Sufficiently Small?

For $\mu > 0$ (2.83) is still valid and $\gamma_1 \neq 0$ also remains true by continuity with respect to μ for μ sufficiently small. Theorem 2.17 guarantees the map $M(\mu)$ is still in the elliptic case, so that the eigenvalues of the linearization of

$M(\mu)$ with respect to the origin still remain on the unit circle. $M(\mu)$ again takes the form (2.84) for μ sufficiently small.

We fix $\mu = \mu^* > 0$ and for this value of μ consider Lemma 2.20. Since $\mu = \mu^*$ is fixed, we suppress its dependence as it will not change. Lemma 2.20 implies that M can be written in the form (2.47), where

$$\lambda(r, \tilde{\mu}) = (2\pi)^{-1}(\gamma_0 + \gamma_1 \tilde{\mu}^2 r), \tag{2.86}$$

$r \in \mathcal{A} = \{0 < r < 1\}, \xi^2 + \eta^2 = \tilde{\mu}^2 r, d\lambda/dr = \tilde{\mu}^2 (2\pi)^{-1}\gamma_1 \neq 0, \tilde{\mu} > 0$, where the terms $\mathcal{O}(\tilde{\mu}^{2k+1}), k = 1$, can be made arbitrarily small as $\tilde{\mu} \to 0$. We now consider $M = M(\tilde{\mu})$. $\tilde{\mu}$ represents the distance to $\xi = \eta = 0$ on $\sum$ (i.e., the distance to $\phi(0, \mu^*)$). Thus, μ^* represents the mass ratio $\frac{m_1}{m_1+m_2}$, while $\tilde{\mu}$ represents the distance to $\phi(0, \mu^*)$. $\lambda(r, \tilde{\mu}) = \lambda(r, \tilde{\mu}, \mu^*)$.

Let $\tilde{\mathcal{A}}$ be a compact subset of $\mathcal{A}$. $\tilde{\mathcal{A}} = \{0 < a \leq r \leq b < 1\}$. Theorem 2.4 can now be applied since M is of the form of (2.47).

Theorem 2.25 *There exists a set $\tilde{\mathcal{C}}$ of invariant closed curves (topological circles) for $M(\tilde{\mu})$ about $\xi = \eta = 0$ provided $\lambda(r)$ satisfies (2.41) for each fixed $\tilde{\mu} > 0$ sufficiently small and $r \in \tilde{\mathcal{A}}$. The induced map of $M(\tilde{\mu})$ on $\tilde{\mathcal{C}}$ is given by (2.40).*

Since we know the dynamics of $M(\tilde{\mu})$ on $\tilde{\mathcal{C}}$, which is quasi-periodic motion with frequency $\lambda(r, \tilde{\mu})$, then *most* of the values of $r \in \mathcal{A}$ will lead to quasi-periodic motion for each fixed $\tilde{\mu}$, for orbits of the restricted problem with initial values lying sufficiently near $\phi(0, \mu^*)$. Let $\tilde{\Lambda}$ represent the set of resonance gaps of $M(\tilde{\mu})$ corresponding to those values of r where (2.41) is not satisfied.

Also, since $\xi^2 + \eta^2 = \tilde{\mu}r$, we can fix $\tilde{\mu}$ sufficiently small since r controls how far to the origin the points lie. The dependence of $\tilde{\mu}$ is now suppressed.

2.5 AUBREY-MATHER SETS AND THE RESTRICTED THREE-BODY PROBLEM

As seen in the previous section M defined by (2.84) can be written as a standard twist map (2.37), where λ is given by (2.86). M is defined in a neighborhood of $\xi = \eta = 0$ corresponding to $\phi(0, \mu^*)$. $(\xi, \eta) = (0, 0)$ is an elliptic fixed point where Theorem 2.17 is satisfied for $k = 1$, i.e., $\gamma_1 \neq 0$. For future reference we refer to an elliptic fixed point satisfying Theorem 2.17 as being of *general elliptic type*.

The goal of this section is to prove some properties of M when restricted to $\tilde{\Lambda}$ which is a set of resonance gaps. One result utilizes the Birkhoff fixed point theorem, where we will deduce the existence of infinitely many periodic points of M near the origin. Another deeper result is to prove the existence of special invariant sets for M called *Aubrey-Mather sets.*

It is remarked that we are using the following *notation:* The map $M = M(\tilde{\mu})$ is constructed for the restricted problem in the previous section with a fixed point of the general elliptic type, and it has a set of invariant curves $\tilde{\mathcal{C}}$ and resonance gaps $\tilde{\Lambda}$, which are guaranteed by the Moser twist theorem. Next we will consider a general area-preserving map $\mathcal{M}$ with a fixed point of the general elliptic type, and its respective family of invariant curves and resonance gaps are labeled as $\mathcal{C}$, Λ, without the $\tilde{}$. The same notation $\mathcal{A}$ is used for the domain of definition of these maps when expressed in the form of a standard twist map (2.37). This is because in both of these cases, since they have fixed points of the general elliptic type, Lemma 2.20 can be applied, yielding the same domain $\mathcal{A}$ where $0 < r < 1, \theta \in [0, 2\pi]$.

We will discuss $\mathcal{M}$ in reference to the Birkhoff fixed point theorem and the Aubrey-Mather theorem, and M is always used for the specific application to the restricted problem. Note that we also used the notation M_μ for a general twist map 2.37 used in the Moser twist theorem. This notation is not used in this section.

2.5.1 The Birkhoff Fixed Point Theorem

The first thing we consider is how to find periodic points for M. They must reside in $\tilde{\Lambda}$ since all the points on $\tilde{\mathcal{C}}$ under iteration by M yield quasi-periodic orbits. The iterations of M of points on each invariant curve are dense on that curve and won't be fixed. By an *orbit* of M of a point $z \in \tilde{\mathcal{C}}$ we mean the set $\{z, M^{\pm 1}z, M^{\pm 2}z, \ldots\}$.

A theorem originally proven by Birkhoff [46] guarantees infinitely many periodic points for M. More generally, it yields infinitely many periodic points for any area-preserving map in a neighborhood of a fixed point of a general elliptic type in the resonance gaps. This represents a fundamental theorem in the subject of the dynamics of area-preserving maps.

More generally, let $\mathcal{M}$ be a smooth area-preserving map (2.42) with fixed point at $x = y = 0$, where the origin is of the general elliptic type. Thus, there exists a smooth change of coordinates where $\mathcal{M}$ can be written in the form (2.45) with ξ, η replaced by x, y, respectively, where $\gamma_1 = \gamma_2 = \gamma_3 = \cdots = \gamma_{k-1} = 0, \gamma_k \neq 0$. It is remarked that $\gamma_j, j = 1, 2, \ldots, \gamma_k$, are invariants for $\mathcal{M}$.

Theorem 2.26 (Birkhoff Fixed Point Theorem) *There exist at least two nonzero fixed points for $\mathcal{M}^n$ in each sufficiently small neighborhood D of the origin in the (x, y)-plane for all sufficiently large positive integers $n > N(D)$, such that $\mathcal{M}^k \mathbf{x} \in D, k = 1, 2, \ldots, n-1, \mathbf{x} = (x, y)$.*

Thus, infinitely many fixed points of $\mathcal{M}^n$, or equivalently, periodic points of $\mathcal{M}$, exist by Theorem 2.26 in each neighborhood of $x = y = 0$. We have stated Theorem 2.26 in the way it is formulated in [204], where the proof is the same as Birkhoff's except that the estimates used in his proof are made more precise. As n varies, there may be overlap in the fixed points belonging to different values of n. This is avoided if n is restricted to the set of prime numbers. The proof of Theorem 2.26 is geometric in nature and uses mathematical induction.

Note that when $\mathcal{M}$ is expressed in Birkhoff normal form (2.45) which in turn can then be expressed in the form of (2.37), $\lambda(r)$ will not satisfy (2.41) for the fixed points given by Theorem 2.26. For these points, $\mathcal{M}$ represents a rational rotation. These periodic points will lie near a circle, since they will lie in the resonance gaps between the invariant closed curves of the Moser twist theorem which are close to circles. λ is rational for periodic points.

We state this more formally as Lemma 2.30, and first make some definitions.

Definition 2.27 *An area-preserving map of the general form of (2.37) is called a* perturbed integrable twist map, *or* standard twist map. $\lambda(r), \lambda' \neq 0$ *is the* rotation number *and (2.41) are called* diophantine or nonresonance conditions. *When $\mu = 0$, (2.37) is called an* integrable twist map. *Equation (2.42) is called a* general area-preserving map. *When (2.41) is not satisfied, λ is called* approximately rational.

Definition 2.28 *An area-preserving map $\mathcal{M}$ written as (2.45) is said to be in* Birkhoff elliptic normal form, *and $\gamma_j, j = 1, 2, \ldots, k, \gamma_1 = \gamma_2 = \cdots = \gamma_{k-1} = 0, \gamma_k \neq 0$, are called* Birkhoff invariants.

Thus, Lemma 2.20 can be stated as follows.

Lemma 2.29 *A general area-preserving map $\mathcal{M}$ with a fixed point of a general elliptic type can be written as a perturbed integrable twist map with rotation number $\lambda(r)$.*

Therefore Theorem 2.26 implies the following lemma.

Lemma 2.30 *An area-preserving map with a fixed point of the general elliptic type at the origin has infinitely many periodic points near the origin given by Theorem 2.26 which exist in the resonance gaps where the rotation number λ of (2.37) is approximately rational.*

Theorem 2.26 provides a way to prove the existence of periodic points in the resonance gaps of a perturbed integrable twist map.

2.5.2 Aubrey-Mather Sets and the Restricted Problem

Applying Theorem 2.26 to M yields interesting generic results for the restricted problem. M satisfies the assumptions of this theorem, with $\gamma_1 \neq 0$, as we have seen. Since n can be made arbitrarily large, this yields infinitely many periodic orbits near $\phi(t, \mu^*)$ with very long periods, cycling many times before returning to their initial values. Since these periodic points of M lie in $\tilde{\Lambda}$, some of them may also be of the general elliptic type. These points will then have their own set of invariant curves surrounding them by Theorem 2.4. Theorem 2.26 can then be applied to those smaller resonance gaps to obtain more periodic orbits. This process can be repeated over and over.

Now, more generally we return to $\mathcal{M}$, where it is satisfies the assumptions for the Birkhoff fixed point theorem. In a neighborhood of the origin, it can be expressed as a perturbed integrable twist map, where $\lambda(r) = (2\pi)^{-1}(\gamma_0 + \gamma_k\tilde{\mu}^{2k}r), x^2 + y^2 < \tilde{\mu}^2, r \in \mathcal{A} = \{(0 < a \leq r \leq b < 1\}, a < b$ and $d\lambda/dr > 0$. $\theta \in \mathbf{R}^1 \mod 2\pi$. $A = \{r, \theta\}$ is an annulus of width $b - a$.

We assume $\tilde{\mu}$ is sufficiently small so that Theorem 2.4 is applicable on A. Therefore $\mathcal{M}$ will have a set $\mathcal{C}$ of invariant curves, where $\lambda(r)$ satisfy nonresonance conditions (2.41).

Let Λ represent the set of resonance gaps for $\mathcal{M}$ where (2.41) is not satisfied. Let $g \in \Lambda$ be a particular gap, which is bounded by two invariant topological circles in $\mathcal{C}$ supporting quasi-periodic motion given by (2.40) on which the iterates of $\mathcal{M}$ are dense.

The work of Mather [146, 145] and Aubrey [17, 18] shed light on the nature of the dynamics of $\mathcal{M}$ in Λ which is assumed to be order preserving (see Definition 2.31). It is referred to as Aubrey-Mather theory. Their work is applicable to a large class of area-preserving twist maps which we define in subsection 2.5.3 in Definition 2.36. Our presentation, which very briefly describes some of this work, is guided by Mather's papers, in particular [146, 145], and also by Katok [121, 119, 120]. Aubrey-Mather theory represents a way to generally describe the dynamics of two-dimensional area-preserving twist maps using variational methods. The maps we are considering, perturbed integrable twist maps (see Definition 2.27), fall in this category. A main result of this approach is the proof of existence of special invariant sets in the plane which can be viewed as generalizations of invariant sets obtained in the theory of homeomorphisms of the circle [120].

As with the KAM theorem (Theorem 2.1 for flows or the Moser twist theorem for maps) Aubrey-Mather theory represents a fundamental area of dynamics. The KAM theorem, although stated here for two-degrees of freedom is valid for n dimensions, $n \geq 2$, whereas Aubrey-Mather theory is valid for two dimensions. Attempts have been made to generalize it [119]. The proof of Theorem 2.4 is analytic in nature where delicate estimates are required to insure the convergence of various infinite series required in the proof, due to so-called *small divisors.* These are expressions in the denominators of the terms of the series which are zero on the set of rational numbers. The proofs of Aubrey and Mather rely on variational methods where the invariant sets represent minimal states. The idea of this method is discussed in subsection 2.5.3.

Consider a neighborhood of $x = y = 0$ for $\mathcal{M}$. Since $\lambda'(r) > 0, \lambda$ is a 1:1 function on $\mathcal{A}$, and we assume $\lambda > 0$. As r varies in $\mathcal{A}, \lambda$ is either rational or irrational. Aubrey-Mather theory proves the existence of special invariant sets depending on whether λ is rational or irrational.

We define what it means for $\mathcal{M}$ to be order preserving on an invariant subset $A_I \subset A$, $\mathcal{M} : A_I \to A_I$. To define this it is convenient not to identify θ mod 2π but rather $\theta \in \mathbf{R}^1$. Thus A becomes an infinite strip S of width $b - a$. The invariant set A_I then corresponds to a set $S_I \subset S$. The infinite strip is called a universal covering space for A and is defined in subsection 2.5.3.

Using S we can define what it means for $\mathcal{M}$ to be order preserving on S_I. Let $\alpha_k = (r_k, \theta_k), k = 1, 2$, be two arbitrary distinct points of S_I. We assume that $\theta_1 < \theta_2$. Let $\mathcal{M}\alpha_k = \alpha_{k+2} = (r_{k+2}, \theta_{k+2})$. Since $\lambda' > 0$, then all points are mapped from left to right on S; $\mathcal{M}$ preserves orientation on S. Therefore $\theta_k < \theta_{k+2}$. The extension of $\mathcal{M}$ to S is called the lift map and is defined in subsection 2.5.3.

Definition 2.31 *$\mathcal{M}$ is order preserving if $\theta_3 < \theta_4$*

Theorem 2.32 (Aubrey, Mather) *There exist three types of invariant sets for $\mathcal{M}$ which are* order preserving *and closed. If $\lambda > 0$ is an irrational number, then the invariant set is topologically equivalent to either, (i) circle or, (ii) Cantor set, on either of which the orbits for $\mathcal{M}$ are quasi-periodic. The Cantor sets are accompanied by order-preserving orbits doubly* *asymptotic to them. If λ is a rational number, then the invariant set is (iii) collection of at least two periodic points of type (p, q) (see subsection 2.5.3), for all relatively prime integers $p > 0, q > 0$ together with homoclinic and heteroclinic orbits.*

We call these order preserving invariant sets (i), (ii), (iii), Types 1, 2, 3, respectively. This is not a general way to refer to these sets in the literature, and is used here only for convenience. The dynamics of $\mathcal{M}$ on the type 1 sets is equivalent to a rotation. The dynamics of $\mathcal{M}$ on the type 2 sets is related to the so-called *Denjoy counterexamples* of circle homeomorphisms [66] and can be viewed, in part, as a generalization of this. The Denjoy counterexamples are maps on the circle which give nontransitive circle diffeomorphisms due to not having sufficiently high differentiability. When a diffeomorphism on the circle has an irrational rotation number and the derivative of the map has at least bounded variation, which is true for continuously differentiable maps, then the diffeomorphism is transitive. This means that it is topologically conjugate to a rotation with this rotation number. The motion is quasi-periodic. Thus, iterates of the map are dense on the circle. If the differentiability is not sufficiently high, then Denjoy found examples of maps that have an irrational rotation number but are not transitive. In the more general context of maps not on a circle but on two dimensions, Mather stated in [145] that the use of the term *quasi-periodic* for the type 2 sets is used in a liberal fashion. That is, the closure of an orbit on a type 2 set is a Cantor set. This definition is an elegant way to extend the definition of quasi-periodic motion when not confined to a circle. The periodic points of the type 3 sets include the periodic points of Theorem 2.26 and are obtained from the more general version of Theorem 2.26 we discuss in Subsection 2.5.3. The invariant sets of types 1, 2, 3 given by Theorem 2.32 are generally referred to as Aubrey-Mather sets.

We comment on the variational methods used in Aubrey-Mather theory in subsection 2.5.3.

Since $\mathcal{M}$ is a perturbed integrable twist map with the origin as a fixed point of the general elliptic type, Theorem 2.4 yields set $\mathcal{C}$ of invariant curves, as we have noted earlier. It could be asked if Theorem 2.32 also yields this set of topological circles supporting quasi-periodic motion. On the set $\mathcal{C}$, λ is irrational and Theorem 2.32 implies an invariant set of type 1 *or* type 2. The type 1 set is equivalent to the invariant curves of $\mathcal{C}$, however, Theorem 2.32 doesn't necessarily guarantee this type. Thus, in this sense Theorem 2.4 yields stronger results in the case of λ being diophantine.

When λ is not diophantine and hence approximately rational we are on the set Λ consisting of resonance gaps. In this case Theorem 2.4 says little, and Theorem 2.32 says much more. When λ is approximately rational, it can be irrational or rational.

One interesting question to ask is if there is any trace of the invariant circles supporting quasi-periodic motion comprising the set C in Λ? That is, in Λ, we know the invariant circles surrounding the origin generically no longer exist, and they have broken up, or as Moser phrases it in [204], they have *disintegrated.* The type 2,3 sets can be viewed as *relics*, or as Katok phrases it, *ghosts* of the invariant circles [119]. For an irrational rotation number λ in a gap $g \in \Lambda$, the type 2 set can be approximated by infinite sequences of periodic points of the type 3 set obtained from a generalized version of Theorem 2.26, which is Theorem 2.35 discussed below. The rational rotation numbers of the periodic points in the sequence converge to the value of λ. The limits of the sequences of periodic points yield a Cantor set. It is not intuitively obvious what this Cantor set looks like. The dynamics on it is generically hyperbolically unstable since generically the type 3 periodic points are hyperbolic. This is a nontrivial construction, and the details are in [146, 120].

The use of the term *generic* with regards to the generic breakup of the invariant topological circles is directly related to the use of the term in Zehnder's theorem, Theorem 2.22. This follows since if a transverse homoclinic point generically exists in the resonance region $\tilde{\Lambda}$ near the elliptic fixed point, then invariant closed curves supporting quasi-periodic dynamics as given by the Moser twist theorem cannot exist in the hyperbolic network given by the transverse homoclinic point. Thus, generically, they break up.

Theorem 2.32 can be applied to the restricted three-body problem. It is directly applicable to M acting on the set $\tilde{\Lambda}$ of resonance gaps near the periodic orbits of the first kind.

Theorem 2.33 *There generically exists Aubrey-Mather sets of types 2 and 3 in $\tilde{\Lambda}$ in every neighborhood sufficiently near the initial condition $\phi(0, \mu)$ of the periodic orbit of the first kind on the section $\sum$, for each μ sufficiently small.*

The term *generically* must be used in the statement of Theorem 2.33. This is because it must be insured that in a subset of $\tilde{\Lambda}$ for the specific map M, invariant closed curves guaranteed by the Moser twist theorem have broken up. As discussed, we know that this will occur generically from Zehnder's theorem. Once we know that they have broken up, then this rules out the type 1 sets.

If we imagine that P_3 represents an asteroid moving about the Sun, P_1, in the asteroid belt between Mars and Jupiter, where P_2 = Jupiter, then Theorem 2.33 gives a sense of the complexity of the motion the asteroid can have in addition to what is shown in Figure 2.14. It is remarked that the condition (2.82) for the frequencies yields a finite set of excluded resonant frequencies corresponding to the so-called *Kirkwood gaps* in the asteroid belt where asteroids are not found [204]. This illustrates the instability of motion in the resonance gaps which is likely related to Arnold diffusion.

Note that type 3 invariant sets of a general perturbed integrable twist map $\mathcal{M}$ have the property that if one considers iterates of $\mathcal{M}$ on the heteroclinic orbits associated to the periodic points of type (p, q), where p, q are relatively prime, then the iterates will move towards one point of type (p, q) to a given neighborhood and then gradually to a given neighborhood of a different point of type (m, n), $m \neq p$ and/or $q \neq n$, n, m relatively prime. This defines a transition from a neighborhood of a (p, q) point to a neighborhood of an (m, n) point. We refer to this as a *resonance transition* $(p, q) \rightarrow (m, n)$. Hence Theorem 2.33 implies the existence of resonance transitions in $\tilde{\Lambda}$ in every neighborhood sufficiently near the initial condition $\phi(0, \mu)$ of the periodic orbit of the first kind on the section $\sum$, for each μ sufficiently small.

Interesting resonance transitions have been observed in comets in orbit about the Sun which rapidly transition between different resonance motions by gravitationally interacting with Jupiter in a complicated fashion. This was studied in [37] using some ideas in chapter 3. This type of resonance transition need not be associated with type 3 sets. The cometary motions analyzed in [37] were later numerically analyzed in [127], which numerically estimated complicated networks of hyperbolic manifolds associated with the transitions. Resonance transitions of this type are studied for Kuiper belt objects in orbit about the Sun which gravitationally interact with Neptune [21, 41].

It is noted that the generality of Theorem 2.32 is as remarkable as its applicability. Of course, to apply it, one has to know that the first Birkhoff invariant $\gamma_1 \neq 0$, which is difficult to prove in general.

2.5.3 (p, q)-Periodic Points, Generalizations

We conclude this section with a brief discussion of periodic points in Λ for $\mathcal{M}$ and generalizations of Theorem 2.32. $\mathcal{M}$ is defined in the previous subsection with the same assumptions. We assume $\alpha \in A$, where $\theta \in \mathbf{R^1} \mod 2\pi$. Let $\alpha = (r, \theta)$ be a periodic point of $\mathcal{M}$. $\mathcal{M}^q \alpha = \alpha$ for some integer $q > 0$. Now, the rotation number $\lambda(r) > 0$ of $\mathcal{M}$ is a rational number Q and can be written as p/q where $p > 0, q > 0$ are integers. That is, it can be expressed up to the integer p. Applying $\mathcal{M}$ q times advances θ by $2\pi p$. That is, after q applications of $\mathcal{M}$, the point α makes p cycles of the origin $r = 0$. This is called a *periodic point of type* (p, q). See Figure 2.16, which shows a periodic point of type (1, 3).

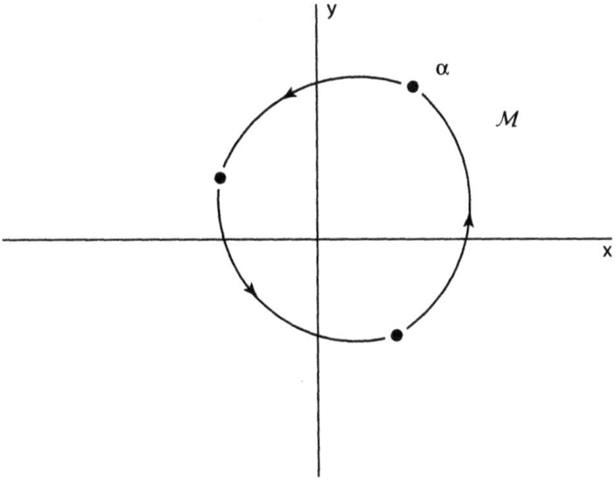

Figure 2.16 Periodic point of type (1,3).

p/q is assumed to be reduced to lowest terms, and p, q are relatively prime, as any integer multiples of p, q lead to the same period point. The iterates of α need not lie on a circle, and the type 3 invariant sets will in general lie near a circle [146, 120].

The classical Birkhoff fixed point theorem, Theorem 2.26, is for the case where $\mathcal{M}$ is smooth. It implies the existence of at least two fixed points for $\mathcal{M}^q$ in a neighborhood of the origin in the (x, y)-plane, $q > N(D)$. In other words, it yields the existence of at least two periodic orbits of type (p, q). It can be proven that these points are order preserving. Thus, Theorem 2.26 can be stated as follows.

Theorem 2.34 *There exist at least two order-preserving periodic points of type* (p, q) *for* $\mathcal{M}, q > N(D), p > 0, q > 0$ *relatively prime.*

This yields infinitely many periodic points as p, q vary over relatively prime numbers for each neighborhood D, that are a subset of the type 3 invariant sets [146]. The existence in Aubrey-Mather theory of the type 3 sets is obtained from a more general version of Theorem 2.26 proven using variational methods. We list this as Theorem 2.35.

Theorem 2.35 *Let* $\mathcal{M}$ *be a diffeomorphism. Then there exist at least two periodic points for each* $\lambda = p/q$ *in the twist interval,* $p > 0, q > 0$ *relatively* *prime. These periodic points are associated to heteroclinic and homoclinic orbits.*

The interval $\lambda(\mathcal{A}) = [\lambda(a), \lambda(b)]$ is called the *twist interval.* $\lambda(r)$ describes the amount to which θ is changed, or twisted, along uniform circles for $\mu = 0$. The magnitude of λ' yields the magnitude of the twist. As p, q vary, infinitely many periodic points are therefore obtained.

We conclude this section with a remark on the generalization of Theorems 2.32 and 2.35 for a more general class of twist maps. Throughout our presentation we have assumed $\mathcal{M}$ was a smooth perturbed integrable twist map. However, this is a special case. Mather's main theorem in [146] applies to area-preserving maps f, which are more general twist maps (defined below) and are homeomorphisms. Theorem 2.35 in this case yields only one periodic orbit of type (p, q) where $\mathcal{M} \equiv f$. If f is assumed to be a diffeomorphism then Theorem 2.35 still yields two periodic points.

We define a general twist map f. In this case it is convenient to work not in the annulus A, but in the universal covering space S. Let

$$A = \{(\theta, r) | \theta \in \mathbf{R^1} \mod 1, r \in [0, 1]\}.$$

Rather than identifying $\theta \mod 1, \theta$ is not identified and we unwind the annulus as shown in Figure 2.17. This defines the covering space S,

$$S = \{(x, y) | x \in \mathbf{R^1}, y \in [0, 1]\},$$

where x corresponds to θ no longer identified, and $y \equiv r$. S is an infinite strip of width 1.

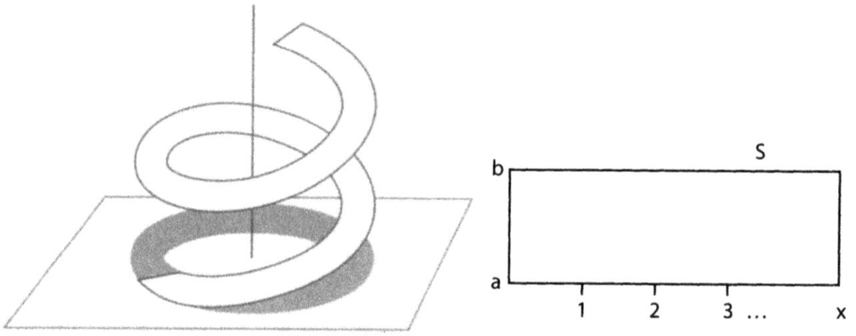

Figure 2.17 Covering space S for lift map.

Let f be a diffeomorphism, $f : A \to A$. Let T be the translation, $T : S \to S, T(x, y) = (x+1, y)$. Viewing f on S yields the *lift* F of f, $FT = TF$, where F is defined up to a power of T. This is f in the coordinates $(x, y) \in S$. In component form

$$F(x, y) = (F_1(x, y), F_2(x, y)).$$

Definition 2.36 $f : A \to A$ *is an area-preserving twist map if it preserves area, preserves boundary components of* A*, preserves orientation, and* $F_1(x,y) > F_1(x,z), y > z$.

Theorems 2.32 and 2.35 are valid for general twist diffeomorphisms f.

We briefly comment on the general idea of the variational method used in the proofs of Theorems 2.32 and 2.35[120]. A function H is defined on the strip S which is a map $H : S \to \mathbf{R}^1_+$, where $\mathbf{R}^1_+$ are the set of real numbers that are nonnegative. Geometrically, H represents the area of a region Q. Q is defined as follows: Consider a vertical line $x = a$ on the strip, where $0 \leq y \leq 1$. We label this line l_a. The lift F maps l_a to a curve γ with a positive slope at each point, where $0 \leq y \leq 1$. Consider a value of x on the x-axis below γ, say at a value $x = b, b > a$. Consider the vertical line $x = b$, l_b, where $0 \leq y \leq 1$. l_b cuts γ into two parts for a value of $y = c \in [0,1]$. There is a part where $x \leq b$, and another where $x \geq b$. The part where $x \leq b$ is on the lower left, and it is defines the region Q. It is bounded above by γ for $x \leq b$ and by the x-axis below. To the right it is bounded by the vertical line l_b. H is the area of this region, and $H = H(a,b)$. H is called the generating function for f. A functional L is defined as a function of H over special sequences of iterates of points in S between two given values of x, and the functional L is minimized over the space of sequences. The minimizing sequences play a key role in the construction of Aubrey-Mather sets. The details of this are beyond the scope of this book and are contained in [120]

Aubrey-Mather theory is an extensive area of study, and we have only touched upon a small aspect of it. Our goal was to apply it to the restricted three-body problem, and not to give a detailed exposition of the theory. This was also our approach when discussing KAM theory.

Chapter Three

Capture

This chapter comprises about one-half of this book, and a number of different topics are covered pertaining to the subject of capture in the three- and four-body problems. Two forms of capture are studied, and in the last section we prove that they are equivalent on a hyperbolic invariant set for the circular restricted problem.

Section 3.1 serves as an introduction to the different types of capture and other dynamics we will consider in this chapter. These include permanent capture and a theorem by Chazy and Hopf on the measure of permanent capture orbits in the general three-body problem. A historical treatment of permanent capture is given from the work of Sitnikov, Alekseev, Moser, and others for the Sitnikov problem. Unbounded oscillatory motion is also defined. Both these types of motions occur near the set of parabolic orbits for the three-dimensional elliptic three-body problem as well as the general n-body problem. In contrast to permanent capture, temporary capture is defined.

Next, in section 3.1 a different type of capture, called ballistic capture is defined for the three-dimensional elliptic problem. A region where this capture can occur in the phase space is called the weak stability boundary and is analytically defined. Ballistic capture occuring on the weak stability boundary is called weak capture. Transfer orbits are defined which lead to ballistic capture. A summary of several different types of capture and other dynamics is listed for reference. It includes notions introduced in later sections such as primary interchange capture and asymptotic capture.

Finally, we end section 3.1 by defining the capture problem in a special restricted four-body problem that is key for the construction of applied ballistic capture transfers.

In section 3.2 the weak stability boundary region is defined for the planar circular restricted problem. It is first defined by a numerical algorithm to motivate its description. However, of interest for this book in later sections is an analytic approximation for this region which is explicitly derived, labeled W.

In section 3.3 the work of Conley is described on the existence of transit orbits in the planar circular restricted problem; these orbits exist in the region connecting the Hill's regions about the Earth and Moon. We prove that transit orbits are ballistic capture transfers. We define asymptotic capture to a periodic orbit based on Conley's work as well as so-called primary interchange capture, which we prove exists. This proof makes use of a result of McGehee. Homoclinic orbits are defined and more general results by Llibre, Simó, and Martinez are breifly discussed. A numerical example illustrating primary interchange ballistic capture is described in the case of three dimensions with more realistic modeling in subsection 3.3.2.

Section 3.4 is an applied section and is the only purely applied section in this book. In it we show that the unstable properties of the weak stability boundary can be used to find a different type of transfer to the Moon using ballistic capture. It has advantages over the classical transfers for useful applications. Such a transfer type was used to successfully bring a Japanese spacecraft to the Moon in 1991. It is constructed in a restricted four-body problem. Its dynamics is briefly described using invariant manifolds and Hill's regions. A straight forward method is described to numerically determine these transfers for applications. In subsection 3.4.3 we describe how the construction of ballistic capture transfers of this type represent a different methodology for the construction of transfers for spacecraft using stability boundaries and other methods of dynamical systems theory.

In section 3.5 we consider parabolic motion in the circular restricted problem. The general existence theorem of parabolic motion by Easton and Robinson is described. The definition of W is extended to include motion which is slightly hyperbolic. We then prove that on the extended set, $\tilde{W}$, the Jacobi energy can achieve a range of values of importance for the last section.

In section 3.6 we first summarize the work of Moser, who gave a simplified proof of the existence of chaotic motion in Sitnikov's problem near the set of parabolic orbits. We summarize the key steps in his proof. He accomplishes this by proving the existence of a hyperbolic invariant set and applying symbolic dynamics. This follows by an extension of the Smale-Birkhoff theorem. We then outline a proof by Xia, who proves an analogous result for the planar circular restricted problem, and show how it follows the methodology of Moser's proof with modifications. He proves the existence of a hyperbolic invariant set Λ near P_2 for $|C| \gtrsim \sqrt{2}$, where permanent capture, unbounded oscillatory motion, and infinitely many other types of motions can be prescribed.

A main result for chapter 3, proved at the end of section 3.6, is that the intersection of Λ with $\tilde{W}$ is nonempty, summarized in Theorem 3.58. This implies that ballistic capture orbits defined on $\tilde{W}$ pass through the set Λ and therefore take on all the chaotic motions associated with that set. This result ties together hyperbolic dynamics on hyperbolic invariant sets with ballistic capture and shows that ballistic capture on $\tilde{W}$ is a chaotic process. Applications are discussed. This will conclude the book.

3.1 INTRODUCTION TO CAPTURE

The basic forms of capture we need for this chapter are defined in this section. In order to consider the problem of capture, the n-body problem is required for $n \geq 3$. In the case of $n = 3$, we will speak of capture of one of the mass particles, say P_3, about one or both of the other two particles. By capture we will mean that P_3 is somehow bound to the other particles. The way P_3 can be bound to the other particles can be defined in different ways. For example, in the elliptic restricted problem, one can imagine that P_3 approaches the other two particles P_1, P_2 from far away, and then is captured about either or both of the particles according to the definition of capture. In the two-body problem the two particles are in only one type of orbit for a given energy and that orbit will not change. This is why $n \geq 3$ is required for a capture to occur from a noncapture state.

Consider the three-dimensional elliptic restricted problem (1.77). This assumes two primary mass particles P_1, P_2 of respective masses m_1, m_2, moving in mutually elliptic orbits about their common center of mass placed at the center of an inertial coordinate system $Q_1, Q_2, Q_3, m_1 = 1 - \mu, m_2 = \mu, 0 < \mu < 1/2$, and moving in the Q_1, Q_2-plane. $e \in [0, 1)$ is the eccentricity of their orbit. The particle P_3 of zero mass, $m_3 = 0$, moves in the gravitational field generated by the mutual elliptic motion of P_1, P_2. For example, P_1 = Sun, P_2 = Jupiter, P_3 = comet; or P_1 = Earth, P_2 = Moon, P_3 = spacecraft, etc. Let $\mathbf{Q} = (Q_1, Q_2, Q_3) \in \mathbf{R}^3$ denote the vector position of P_3 (see Figure 3.1).

3.1.1 Capture and Bounded Motion

The first type of capture we define is of a global type and is termed *total capture* or alternatively *permanent capture.*

Definition 3.1 P_3 *is* permanently captured *into the* P_1, P_2*-system in* forward time *if as* $t \to \infty, |\mathbf{Q}|$ *is bounded, and as* $t \to -\infty, |\mathbf{Q}| \to \infty$*.* P_3 *is*

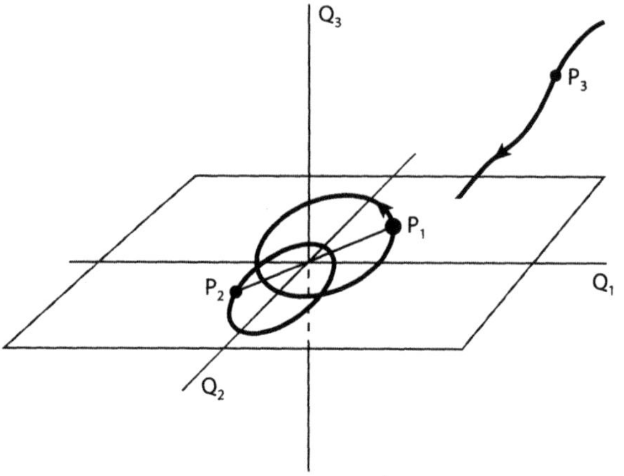

Figure 3.1 Three-dimensional elliptic restricted three-body problem.

permanently captured in backward time *if as* $t \to -\infty, |\mathbf{Q}|$ *is bounded and as* $t \to \infty, |\mathbf{Q}| \to \infty$.

More generally, this definition is extended to the full three-dimensional n-body problem for mass particle $m_k, k = 1, 2, \ldots, n$, defined by (1.1), $n > 3$.

Definition 3.2 *Permanent capture for the three-dimensional n-body problem occurs in forward time if as* $t \to \infty$, *all* $r_{k\ell}, k > \ell$, *are bounded, and as* $t \to -\infty$, *at least one of the distances* $r_{k\ell}$ *tends to infinity. An analogous definition is given for permanent capture in backward time.*

We consider $n = 3$. Permanent capture is difficult to achieve, and it was originally proven by Chazy [54, 55] that the set of initial values leading to it comprise a *set of measure zero*; Hopf gave another proof [114, 204]. A natural volume element for the measure in this problem is given by $d\mu = dx_{11}dx_{12}dx_{13}dx_{21}\ldots dx_{33}d\dot{x}_{11}d\dot{x}_{12}\ldots d\dot{x}_{33}$. The volume of phase space with points leading to permanent capture is zero, and permanent capture is unlikely to exist, if it exists at all.

Theorem 3.3 (Chazy) *The set of orbits leading to permanent capture in the general three-body problem is a set of measure zero.*

Theorem 3.3 *does not* imply that permanent capture orbits necessarily exist. This is a much more difficult and deeper question and was first solved by Sitnikov for a special version of the three-body problem, now referred to as the *Sitnikov problem*, described below. Alekseev later proved this by more sophisticated methods. He generalized his proof to a more general version of Sitnikov's problem, and then to the general three-body problem. Let's look at this more carefully.

The Sitnikov problem is a special version of the elliptic restricted problem shown in Figure 3.1. In it, P_3 is constrained to lie on the Q_3-axis and $\mu = 1/2$. The Q_3-axis is verified to be a two-dimensional invariant submanifold with coordinates $Q_3, \dot{Q}_3$. That is, setting $Q_1 = 0, Q_2 = 0, \mu = 1/2$ in (1.77) yields a single second order differential equation in the variable Q_3 for P_3,

$$\ddot{Q}_3 = -\frac{Q_3}{(Q_3^2 + r^2(t))^{\frac{3}{2}}}, \tag{3.1}$$

$r(t) = (Q_1^2(t) + Q_2^2(t))^{\frac{1}{2}}; r(t)$ is given by (1.47). In this simplified restricted problem, the difficulties of achieving permanent capture can be seen. For example, if an initial value $Q_3(0), \dot{Q}_3(0)$ is given where $|\dot{Q}_3(0)|$ is sufficiently large, then it can be shown that P_3 will escape to infinity, that is, $|Q_3(t)| \to \infty, t \to \infty$. On the other hand, if $|Q_3(0)| + |\dot{Q}_3(0)|$ is sufficiently small, it can be shown that $Q_3(t)$ is bounded for all t. There are critical values of $Q_3(0), \dot{Q}_3(0)$ where P_3 will lie between bounded and unbounded motion. These critical values correspond to escaping orbits which arrive at infinity with zero velocity. That is, as $t \to \infty$, $|Q_3(t)| = \infty$, and $\dot{Q}_3(t) \to 0$.

These are called *parabolic orbits* and can be viewed as critical escape orbits. An escape orbit in (1.77) satisfies $\lim_{t\to\infty} |\mathbf{Q}| = \infty$. Parabolic orbits divide the orbit space into those which escape to infinity where $\lim_{t\to\infty} |\dot{Q}_3(t)| > 0$, and those which remain bounded for all time.

Permanent capture orbits therefore lie at the boundary between hyperbolic and bounded orbits. Clearly, this region may give rise to very sensitive orbits which can perform complicated motions. Sitnikov [208] first proved that permanent capture exists in (3.1). Alekseev [6] proved this by different methods. The motion was proven by Alekseev to be chaotic. This type of chaotic motion is described in section 3.6 and results from a hyperbolic invariant set.

Intuitively, it is not too difficult to see that permanent capture can occur. Since P_3 is between bounded and hyperbolic motions, a tiny velocity increase can cause P_3 to become unbounded in backwards time, whereas a tiny velocity decrease can cause P_3 can become bounded for all forward time. Of course, proving this is difficult.

Another interesting motion exists between hyperbolic and bounded motions. In this case $Q_3(t)$ becomes *unbounded but does not tend to infinity.* That is, as $t \to \pm\infty, \lim |Q_3(t)|$ does not exist. The particle oscillates on the Q_3-axis where the amplitudes of the oscillations increase as $t \to \infty$. P_3 moves in the positive Q_3-axis to a maximal distance, then falls back, passing through the $Q_3 = 0$ plane and down the minus Q_3-axis to a point where $|Q_3|$ is larger than the previous maximum of $|Q_3|$. P_3 then moves up the Q_3-axis, passing again through $Q_3 = 0$, and up to a value of $|Q_3|$ larger than the previous maximum. The consecutive times of passage $t_{k+1} - t_k$ through $Q_3 = 0$ tends to infinity. Thus, limsup of $|Q_3(t)| = \infty$, but $\liminf |Q_3(t)| = 0$.

This type of motion was first hypothesized by Chazy. It is called *oscillatory motion*, first proven to exist by Alekseev [7], and described in section 3.6. Alekseev generalized his proof of both oscillatory motion and permanent capture in the Sitnikov problem to the case where $m_3 > 0$. He also proved similar results in the general three-body problem for negative energy where the mass values are restricted.

Alekseev's methods are lengthy. Moser gives a more simplified proof of the existence of complicated dynamics near parabolic orbits in Sitnikov's problem [175]. This is outlined in subsection 3.6.4. It uses a transformation of McGehee, which shows that the set of parabolic orbits is asymptotic to an unstable periodic orbit at $|Q_3| = \infty$. The two-dimensional set of parabolic orbits going to $|Q_3| = \infty$ corresponds to the stable manifold of this periodic orbit. This is discussed in sections 3.5 and 3.6. A transformation is defined in a neighborhood of this periodic orbit, and it is shown by Moser to give rise to a complicated dynamics.

The methodology of proof by Moser for the chaotic motions near the parabolic orbits in Sitnikov's problem is used by Llibre and Simó [137] to prove analogous results for the planar circular restricted problem. Their proof requires that the Jacobi constant C is sufficiently large and μ is sufficiently small. That is, $C \gg 1, \mu \ll 1$. This implies that P_3 moves far from the two primaries for all time.

We are interested in the case where P_3 passes very near to P_2 repeatably as it is permanently captured. This is not satisfied in [137]. This means the Jacobi constant needs to be suitably restricted. Since P_3 is moving near parabolic orbits while it is permanently captured, this implies for the initial conditions that interests us, $C \approx \pm\sqrt{2}$ in the scaled coordinates used in section 3.5. This value turns out to be used by Xia [227] in his proof of chaotic motions occurring near parabolic orbits in the restricted problem for similar initial conditions for $\mu \ll 1$. Xia also proves the existence of chaotic motions in the case of $C \gg 1$ for all but a finite set of $\mu \in (0,1)$, extending the earlier results by Llibre and Simó where $\mu \ll 1$. He uses the methodology of Moser's proof as well as a basic theorem of Easton and Robinson on the existence of parabolic orbits, which we state in section 3.5. Xia's proof is outlined in subsection 3.6.5. He proves, in particular, the existence of a hyperbolic invariant set Λ for an area-preserving map on a two-dimensional surface of section to the flow near the set of parabolic orbits which pass by P_2. The dynamics of the orbits passing through points of this set leads to a very complicated chaotic motion. It turns out that Λ intersects another set that we will describe later in this section, and in more detail in Section 3.2.

Precise definitions of oscillating motion and parabolic orbits are now given.

Definition 3.4 Unbounded oscillatory motion *for the general n-body problem (1.1) occurs when*

$$\limsup_{t\to\pm\infty} r_{jk} = \infty, \qquad \liminf_{t\to\pm\infty} r_{jk} < \infty \tag{3.2}$$

for at least one pair of masses, $n \geq 3, j < k$.

The term *oscillatory motion* often used in the literature has been modified to include the term *unbounded* here. This makes the definition more precise since the word oscillatory in general includes the idea of bounded motion.

Definition 3.5 Unbounded oscillatory motion *for the three-dimensional elliptic problem satisfies*

$$\limsup_{t\to\pm\infty} |\mathbf{Q}(t)| = \infty, \qquad \liminf_{t\to\pm\infty} |\mathbf{Q}(t)| < \infty. \tag{3.3}$$

As $|t|$ increases, P_3 oscillates so that $|\mathbf{Q}|$ goes out to a maximum, then decreases; then the maxima increases and the oscillations go further and further out. The number of oscillations increases to infinity as $|t| \to \infty$. P_3 is forever oscillating and moving further from the binary pair. Meanwhile, the set of minima of $|\mathbf{Q}(t)|$ are not tending to infinity.

As an example, one could imagine a comet moving about the Sun whose periapsis distances with respect to the Sun are bounded and whose apoapsis distances are far beyond the orbit of Pluto and steadily increasing over each periodic cycle of the comet as times increases.

We know that the set of points leading to permanent capture is a set of measure zero. On the other hand, the measure of the set of points leading to unbounded oscillatory motion is not known.

Definition 3.6 *A parabolic orbit for the elliptic restricted problem satisfies*

$$\lim_{t\to\pm\infty} |\mathbf{Q}(t)| = \infty, \qquad \lim_{t\to\pm\infty} |\dot{\mathbf{Q}}(t)| = 0. \tag{3.4}$$

The case $t \to +\infty$ *defines* ω-parabolic *orbits, and* $t \to -\infty$ *defines* α-parabolic *orbits.*

With a special change of coordinates due to McGehee, we show in section 3.6, that the manifold of parabolic orbits separates elliptic and hyperbolic motion when $\mu = 0$. It is also shown that the parabolic state $|\mathbf{Q}| = \infty, |\dot{\mathbf{Q}}| = 0$ can be mapped into an unstable periodic orbit with two-dimensional stable and unstable manifolds, representing the parabolic orbits.

One can view a parabolic orbit as an escape orbit with minimal energy. More energetic escape orbits are given by hyperbolic orbits.

Definition 3.7 *A hyperbolic orbit for the elliptic restricted problem satisfies*

$$\lim_{t \to \pm\infty} |\mathbf{Q}(t)| = \infty, \qquad \lim_{t \to \pm\infty} |\dot{\mathbf{Q}}(t)| > 0. \tag{3.5}$$

Definition 3.8 P_3 *is* ejected, *or alternatively* escapes *the* P_1, P_2*-system in the elliptic restricted problem in* forward time *if*

$$\lim_{t \to \infty} |\mathbf{Q}(t)| = \infty.$$

This is referred to as unbounded escape *(see Definition 3.14 for* ballistic escape*). If* P_3 *goes beyond a given finite distance* ρ *at a time* t, $|\mathbf{Q}(t)| > \rho$, *then* P_3 *has a* bounded escape.

There is an analogous definition for ejection, or escape, in backwards time.

Hyperbolic and parabolic orbits both lead to bounded and unbounded escape. An orbit in permanent capture in forward time leads to ejection in backwards time.

Finally, we have the following definition.

Definition 3.9 P_3 *has* temporary capture *at* $t = t_0, |t_0| < \infty$, *if* $|\mathbf{Q}(t_0)| < \infty$, *and*

$$\lim_{t \to \pm\infty} |\mathbf{Q}(t)| = \infty.$$

Temporary capture means that P_3 is not permanently captured and is not an unbounded oscillatory orbit. It is ejected as $t \to \pm\infty$.

This definition of temporary capture can be modified to *finite temporary capture* relative to $|Q(t)|$ achieving a prescribed distance $|Q(t)| = d > 0$ at time T. In this case $\Delta = |T - t_0| > 0$ represents the duration of the capture relative to a reference time t_0. This definition of finite temporary capture measures the time that P_3 remains within a given neigborhood of P_2. This definition says little of the dynamical properties of the motion.

3.1.2 Capture Analytically Defined

Consider the elliptic restricted problem. We now consider a different way to define capture. Permanent capture is a geometric notion and requires that P_3 be bounded for positive or negative time. This boundedness is with respect to physical space. Analogously, ejection is also defined in a geometric way. That is, P_3 need only become unbounded.

We distinguish this previous type of capture defined *geometrically* from another type of capture we define *analytically*, called *ballistic capture.*

Ballistic capture is analytically defined for the elliptic restricted problem, and it monitors the sign of the Kepler energy function with respect to the smaller primary P_2. The definition of ballistic capture is generalized in the natural way for the general n-body problem, and that is left as an exercise.

Let X_1, X_2, X_3 be an inertial coordinate system for the elliptic restricted problem *centered at* P_2.

Definition 3.10 *The two-body Kepler energy of P_3 with respect to P_2 in P_2-centered inertial coordinates is given by*

$$E_2(\mathbf{X}, \dot{\mathbf{X}}) = \frac{1}{2}|\dot{\mathbf{X}}|^2 - \frac{\mu}{r_{23}}, \tag{3.6}$$

where $r_{23} = |\mathbf{X}|, 0 \le \mu < 1/2$.

Definition 3.11 P_3 *is* ballistically captured *at P_2 at time $t = t_1$ if*

$$E_2(\varphi(t_1)) \le 0 \tag{3.7}$$

for a solution $\varphi(t) = (\mathbf{X}(t), \dot{\mathbf{X}}(t))$ of the elliptic restricted problem relative to P_2, $r_{23}(\varphi(t)) > 0$.

In particular, in section 3.2 we consider the planar circular restricted problem and determine the set on $\tilde{J}^{-1}(C)$ where $\tilde{E}_2(\mathbf{x}, \dot{\mathbf{x}}) \le 0$ and $\mathbf{x} = (x_1, x_2)$ are barycentric rotating coordinates (1.59). $\tilde{E}_2(\mathbf{x}, \dot{\mathbf{x}})$ is $E_2(\mathbf{X}, \dot{\mathbf{X}})$ expressed in the barycentric rotating coordinates. In addition, those points are considered where $\dot{r}_{23} = 0$. This last condition defines points where solutions are locally normal to radial lines from P_2, i.e., local periapsis or apoapsis points. Set

$$\Sigma = \{\mathbf{x}, \dot{\mathbf{x}} | \tilde{E}_2 \le 0\}, \qquad \sigma = \{\mathbf{x}, \dot{\mathbf{x}} | \dot{r}_{23} = 0\}. \tag{3.8}$$

Then

$$W = \tilde{J}^{-1}(C) \cap \Sigma \cap \sigma \tag{3.9}$$

defines a special set where ballistic capture occurs in the restricted problem [32].

Definition 3.12 *W is called the* weak stability boundary.

W is a set with interesting applications that we discuss in this chapter. It is derived in section 3.2, and C must be suitably restricted. The motion of P_3 near W is sensitive.

In section 3.5 the definition of W is extended to the case where $E_2 \gtrsim 0$ so as to include slight hyperbolic two-body motion with respect to P_2, with $\mu \ll 1$. A point with respect to P_2 satisfying $E_2 \gtrsim 0$ is said to be in *pseudoballistic capture*. The notation $a \gtrsim b$ means that $a - b = \delta > 0$, $\delta \ll 1$. That is, a is strictly slightly greater than b. If $a - b = \delta \geq 0$, $\delta \ll 1$, then we use the notation $a \gtrapprox b$. These definitions extend in a natural way to $<$. It is recalled that $a \ll 1$ means that a is arbitrarily close to 0, where $a > 0$. $a \gg b$ means that $(a-b)^{-1} \ll 1$.

Another extension of W that we make in section 3.5 is to include points where $\dot{r}_{23} \geq 0$. The set W with the two extensions $E_2 \gtrsim 0$, $|\dot{r}_{23}| \geq 0$, we denote by $\tilde{W}$.

A result 3 we prove in section 3.6 is that

$$\tilde{W} \cap \Lambda \neq \emptyset, \tag{3.10}$$

where Λ is a hyperbolic network defined in the beginning of this chapter associated to a transversal homoclinic point, and $\mu \ll 1$. This network is associated to orbits that are near parabolic and is stated more precisely in Theorem 3.58 in subsection 3.6.6. This implies, among other things, that there exist points on $\tilde{W}$ that lead to both permanent capture and unbounded oscillatory motion. This result connects the concept of ballistic capture on $\tilde{W}$ to permanent capture and moreover proves the existence of a hyperbolic invariant set on $\tilde{W}$ giving rise to chaotic motion that is valid for $C \approx \pm\sqrt{2}$. For this motion, P_3 passes near to P_2 and is close to the set of parabolic orbits. The complexity of Λ is equivalent to that shown in Figure 2.8. The part of $\tilde{W}$ that intersects Λ is the one where $E_2 \gtrsim 0$.

It is important to note that the case of interest is when ballistic capture occurs on $\tilde{W}$. We refer to this as *weak ballistic capture.* As will be seen, $\tilde{W}$ gives rise to unstable motion. So, ballistic capture occurring on this set is not stable. We also previously defined ballistic capture in a more general way where it need not occur on $\tilde{W}$. It will be clear from the discussion whether or not ballistic capture is occurring on $\tilde{W}$, and so the word *weak* will not be used for brevity.

Of applied interest are trajectories that go to ballistic capture. This occurs when a trajectory that starts near to the primary P_1 goes to a point on $\tilde{W}$ near P_2. A trajectory of this type called a *ballistic capture transfer* and it has the property that it arrives at a periapsis point near P_2 with substantially lower Kepler energy E_2 than the classical transfer trajectories, called Hohmann transfers and used in applications. This is proven in section 3.3, where a ballistic capture transfer that was actually used to bring a spacecraft from the Earth to the Moon is described. These transfers are of practical importance since for spacecraft they require substantially less fuel to slow down in order to go into orbit about P_2. In section 3.4 we discuss the benefits of this property in the field of aerospace engineering.

Let $\varphi(t)$ be a smooth solution to the elliptic restricted problem for $t_1 \leq t \leq t_2, t_1 < t_2$; t_2 is finite.

Definition 3.13 *If $E_2(\varphi(t_2)) \leq 0$ then $\varphi(t)$ is called a* ballistic capture transfer *from $t = t_1$ to $t = t_2$, relative to P_2.*

It is remarked that E_2 need not be a monotone decreasing function and it could even take on negative or zero values for $t \leq t_2$.

Definition 3.14 *If $E_2(\varphi(t_1)) \leq 0$, and $E_2(\varphi(t_2)) > 0$ then $\varphi(t)$ is called a* ballistic ejection transfer *from $t = t_1$ to $t = t_2$, which defines* ballistic ejection (or escape) *from P_2.*

Let $\varphi(t)$ be a ballistic capture transfer from t_1 to t_2.

We distinguish two types of ballistic capture transfers.

Definition 3.15 *If $t_1 = -\infty$ and $\lim_{t \downarrow -\infty} |\varphi(t)| = \infty$ in Definition 3.13, where $\lim_{t \downarrow -\infty} E_2(\varphi(t)) > 0, E_2(\varphi(t_2)) \leq 0, t_2$ finite, the $\varphi(t)$ is called an* unbounded *ballistic capture transfer. (If $|t_1| < \infty$, then $\varphi(t)$ is a* bounded *ballistic capture transfer as in Definition 3.14.)*

Thus, for an unbounded ballistic capture transfer, P_3 starts infinitely far from P_2 and moves to ballistic capture at P_2 at time $t = t_2$. This would imply permanent capture if $|\varphi(t)| < \infty$ for all $t \geq t_2$.

Analogous to Definitions 3.1 and 3.9, we now define temporary ballistic capture.

Definition 3.16 *Let $\varphi(t)$ be a ballistic capture transfer from $t = t_1$ to $t = t_2$ as in Definition 3.13. If $E_2(\varphi(t)) \leq 0$ for $t_2 \leq t \leq t_3, t_2 < t_3 < \infty$,* *and* $E_2(\varphi(t)) > 0$ *for* $t > t_3$*, then* $\varphi(t)$ *has* temporary ballistic capture *for* $t_2 \le t \le t_3$. *If* $t_3 = \infty$*, then* $\varphi(t)$ *has* permanent ballistic capture *for* $t_2 \le t \le \infty$. *(See Figure 3.2.)*

Definition 3.17 *If the condition* $E_2(\varphi(t_2)) \le 0$ *is replaced by* $E_2(\varphi(t)) \gtrsim 0$ *in Definition 3.13, then* $\varphi(t_2)$ *has pseudoballistic capture with respect to* P_2 *at* $t = t_2$ *and* $\varphi(t)$ *is called a* pseudoballistic capture transfer *from* $t = t_1$ *to* $t = t_2$.

Permanent ballistic capture does not imply permanent capture since if $\varphi(t)$ is permanently ballistically captured for $t \ge t_2$, it need not be unbounded as $t \to -\infty$, and as $t \to \infty$, $\varphi(t)$ need not be bounded.

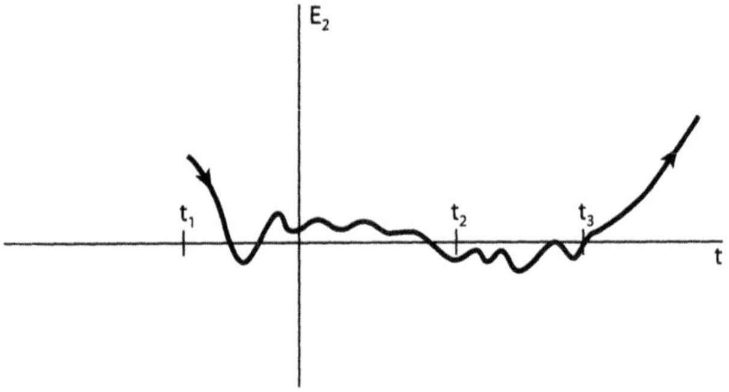

Figure 3.2 Temporary ballistic capture.

If $\varphi(t)$ is permanently ballistically captured for $t \ge t_2$ and $|\varphi| < \infty, t \ge t_2$, and $\lim_{t \downarrow -\infty} |\varphi(t)| = \infty$, then $\varphi(t)$ is permanently captured.

In Section 3.6 we will see how permanent capture leads to an infinite sequence of temporary pseudoballistic captures.

In Table 3.1 we list the various types of capture and transfers defined in this section and in later sections, for reference. The classification presented here by no means includes all possibilities, only those relevant to our discussion.

3.1.3 The Capture Problem

We now turn our attention to the definition of the *capture problem* which is defined for application to the motion of spacecraft. It yields as a solution a

Table 3.1 Captures, escapes, transfers, and orbit types with definition number.

Capture		Capture Transfers	
Permanent	3.1	Ballistic	3.13
Temporary	3.9	Bounded Ballistic	3.15
Ballistic	3.11	Unbounded Ballistic	3.15
Temporary Ballistic	3.16	Pseudo-Ballistic	3.17
Pseudo-Ballistic	3.17	Primary Interchange	3.20
Primary Interchange	3.20		
Asymptotic	3.31		

Escape		Orbit Types	
Bounded	3.8	Unbounded Oscillatory	3.5
Unbounded	3.8	Parabolic	3.6
Ballistic	3.14	Hyperbolic	3.6
Primary Interchange	3.19		

special weak ballistic capture transfer which is not only of practical interest but also of theoretical interest.

A special restricted four-body problem is defined. We first consider a planar elliptic restricted three-body problem between the particles P_1, P_2, P_3. The coordinate system Q_1, Q_2 is inertial and centered at P_1 at the origin. The mass of P_1 is $m_1 > 0$; the mass of P_2 is $m_2 > 0$, where $m_2 \ll m_1$; and the mass of P_3 is zero. All the particles move in the Q_1, Q_2-plane, and P_1, P_2 describe mutual uniform Keplerian elliptic motion about their common center of mass of eccentricity $e_{12} \approx 0$. A fourth mass point P_4, of mass $m_4 > 0$, is introduced. It is assumed to move about the center of mass point P_{cm} between P_1, P_2 in a uniform Keplerian ellipse of eccentricity $e_{124} \approx 0$. Since $m_2 \ll m_1$, then $P_{cm} \approx (0,0)$ and P_4 approximately moves about P_1. We assume that the distance of P_4 from the center of mass of P_1, P_2 is much larger than the distance between P_1 and P_2, and that $m_1 \ll m_4$. The zero mass particle P_3 moves in the gravitational field generated by the assumed elliptic motions of P_1, P_2, P_4.

We refer to this model as a *planar elliptic restricted four-body problem*, which we label as ER4BP-2D. It is shown in Figure 3.3. It could also be referred to as the *co-elliptic restricted four-body problem*. An example of this type of problem is where P_1=Earth, P_2=Moon, P_3=spacecraft, P_4=Sun. We will sometimes refer to P_k with this labeling for convenience without loss of generality. If $e_{12} = e_{124} = 0$, then we refer to this as a co-circular restricted four-body problem. When $m_1 = 1 - \mu, \quad m_2 = \mu, \quad \mu = m_2/(m_1 + m_2) \ll 1, \quad m_4 = 0, \quad e_{12} = 0$, then this problem reduces to the planar circular restricted problem between P_1, P_2, P_3. When $m_4 = 1 - \mu, \quad m_2 = 0, \quad m_1 = \mu, \quad \mu = m_1/(m_1 + m_4) \ll 1, \quad e_{124} = 0$, then the circular restricted problem is obtained between P_1, P_3, P_4. These two restricted problems are considered in subsection 3.4.2.

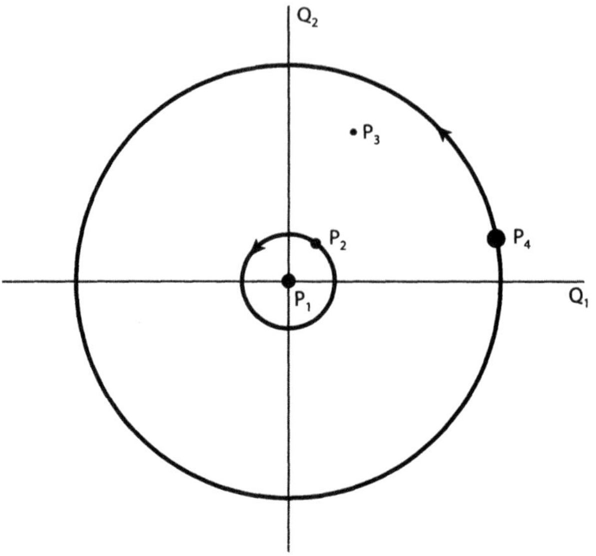

Figure 3.3 A restricted four-body problem.

We define transfers from P_1 to P_2. In Figure 3.4 we just show P_1, P_2, which shows the conditions required for a transfer of P_3 from P_1 to P_2. P_4 is not shown. ER4BP-2D modeling is assumed.

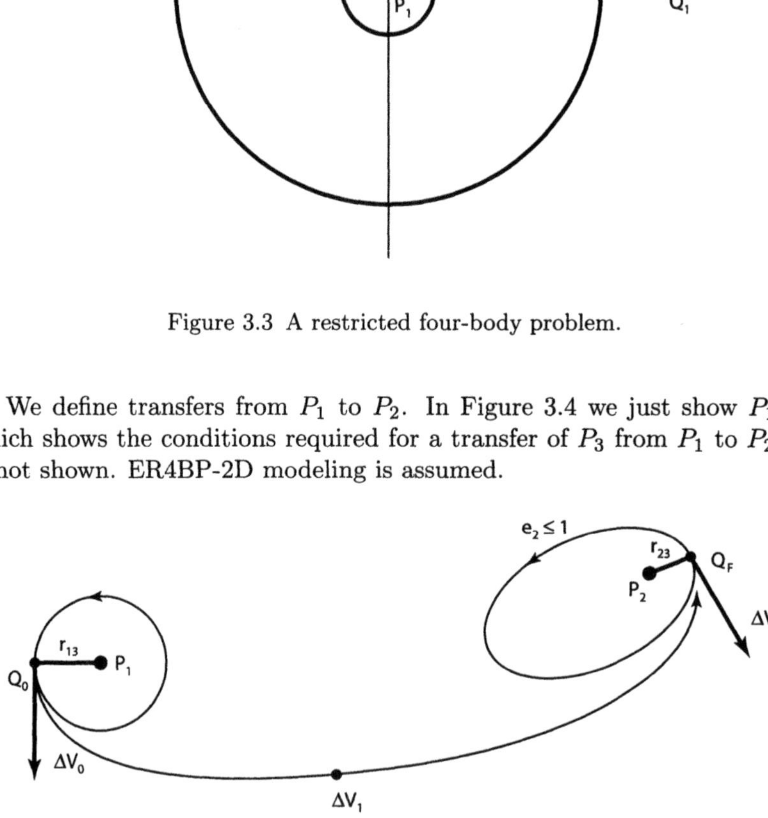

Figure 3.4 The capture problem in inertial coordinates centered at P_1.

The following assumptions are made:

- A1: P_3 initially moves in a circular orbit about P_1 of radius r_{13}.
- A2: A velocity increment magnitude ΔV_0 at the location $\mathbf{Q}_0$ at time t_0 on the circular orbit is added to the circular velocity $(Gm_1r_{13}^{-1})^{\frac{1}{2}}$ so P_3 can transfer to the location $\mathbf{Q}_F$ near P_2. Note that the vectors

$\mathbf{Q}_0, \mathbf{Q}_F$ are in the coordinate system Q_1, Q_2, centered at the origin, P_1.

- A3: A velocity increment magnitude ΔV_1 is applied at a time $t_1, t_0 < t_1 < t_F; t_F$ is the arrival time at $\mathbf{Q}_F$.

- A4: A velocity increment magnitude ΔV_C is applied at $\mathbf{Q}_F$ in order that the two-body energy E_2 between P_2, P_3 be negative or zero at $\mathbf{Q}_F$ so that at $t = t_F$ an oscillating ellipse of given eccentricity $0 \leq e_2 \leq 1$ is obtained of periapsis distance r_{23}. This defines an *instantaneous capture* at $\mathbf{Q}_F$ at time $t = t_F > t_0$ into an ellipse or parabola. We regard a parabola as an ellipse of infinite semimajor axis. $\mathbf{Q}_F$ represents the periapsis of the oscillating ellipse with respect to P_2 at distance r_{23}.

Now we make a few remarks.

3.1.4 Remarks

Remark 1. The term *instantaneous capture* is used in A4 even though $E_2 \leq 0$. The term ballistic capture is not used here since achieving $E_2 \leq 0$ is due to a velocity increment ΔV_C being applied at $\mathbf{Q}_F$. If no velocity increment is necessary and the transfer naturally achieves $E_2 \leq 0$ at $\mathbf{Q}_F$, then that defines ballistic capture, and it is then the same as instantaneous capture.

Remark 2. The velocity increments $\Delta V_0, \Delta V_1, \Delta V_C$ are provided by firing the rocket engines of the spacecraft to impart a thrust, and hence a change in velocity. These increments are called *ΔV's* or *maneuvers.* In practice they cannot be achieved instantaneously, as we are assuming here, and depend on the magnitude of the ΔV. The engines may need to fire for a duration of a few seconds or several minutes. In general, modeling the ΔV's in an instantaneous or *impulsive* manner yields accurate modeling.

Remark 3. The term instantaneous capture in A4 also implies that for $t > t_F$ E_2 may become positive again. That is, P_3 is ejected right after being captured. As we will see, it is generally the case that for ballistic capture, the motion of P_3 for $t > t_F$ is very unstable and ejection generally occurs soon after t_F. Thus, the ellipse shown in Figure 3.4 about the Moon may just exist when $t = t_F$. If ΔV_C is sufficiently large, then the capture ellipses can be stabilized for long times after $t = t_F$. In general if it is desired to place a spacecraft several hundred kilometers from the surface of the Moon in a circular orbit after applying ΔV_C, then the orbit remains approximately circular for several months. Frequent ΔV's need to be applied by the spacecraft to maintain an orbit about the Moon; in general these are not stable due to nonuniformities of the mass distribution of the Moon and gravitational perturbations due to the Earth.

Remark 4. The term *osculating ellipse* in A4 means that the elliptical state at $t = t_F$ at $\mathbf{Q}_F$ may be unstable.

Definition 3.18 *The* capture problem *is defined by the problem*

$$\min\{\Delta V_0 + \Delta V_1 + \Delta V_C\}, \tag{3.11}$$

where the minimization is taken over all transfers from $\mathbf{Q}_0$ *to* $\mathbf{Q}_F$ *and for assumptions A1–A4.*

The capture problem formulated as it is in Definition 3.18 led to the formulation of the weak stability boundary in section 3.2 and to a numerically generated ballistic capture transfer to the Moon of practical interest described in section 3.3. However, another much more interesting ballistic capture transfer to the Moon was operationally tested and is the focus of section 3.4, from both an applied and a theoretical point of view.

Up until the useful demonstration of ballistic capture transfers, the way lunar transfers generally were calculated utilized a vastly more simplified modeling, and at $\mathbf{Q}_F$ have the property that $E_2 \gg 0, \Delta V_C \gg 0$; i.e., they are substantially hyperbolic with respect to the Moon at $\mathbf{Q}_F$. This implies that a substantially larger value of ΔV_C needs to be used relative to ballistic capture. These transfers are called *Hohmann transfers*, and for perspective we will briefly describe them relative to the capture problem. It will be seen in section 3.3 that for a set of given values of $e_2 \leq 1, \Delta V_C = 0$ is achieved for a class of ballistic capture transfers. It is this property that distinguishes them from previous lunar transfers of the Hohmann transfer class.

We conclude this section with a brief description of the Hohmann transfer only for the sake of historical interest and for comparison to the ballistic capture transfer. The Hohmann transfer will not be described further in the book. There are many references to it in the astrodynamics literature [20, 118].

3.1.5 Hohmann Transfer

The Hohmann transfer was developed by W. Hohmann in the early 1900s [110]. Although his assumptions are oversimplifying in nature, they nevertheless lead to transfers from $\mathbf{Q}_0$ to $\mathbf{Q}_F$ which are very useful in practice, not just for the case P_1 = Earth, P_2 = Moon, P_4 = Sun, but for transfers from the Earth to the other planets of our solar system. Also, the Hohmann transfer can be viewed more generally as a way to transfer from a circular orbit about primary mass to another point in space, which need not be near a secondary body.

Here, we discuss the Hohmann transfer that is relevant to Figure 3.4 for the Earth–Moon system for the sake of argument, which can be generalized accordingly to other situations. We assume that P_1, P_2 are in mutually circular orbits.

The basic assumptions in [110] are the following: First, $m_4 = 0$ so that P_4 is not considered. Second, as P_3 transfers from $\mathbf{Q}_0$ to $\mathbf{Q}_F$, i.e., for $t_0 \leq t \leq t_F$, the gravity of P_2 is ignored, i.e., $m_2 = 0$. This yields a simple two-body problem between P_3, P_1, where then the two-body energy is then minimized. This gives one-half of a Kepler ellipse with periapsis at $\mathbf{Q}_0$ and apoapsis at $\mathbf{Q}_F$. This is the Hohmann transfer from $\mathbf{Q}_0$ to $\mathbf{Q}_F$. This ellipse arc has an eccentricity e_1. Upon arrival at $\mathbf{Q}_F$, the gravity of m_1 is ignored, $m_1 = 0$. m_2 is now assumed to be nonzero, and ΔV_C is computed relative to a two-body problem between P_3, P_2. $\mathbf{Q}_F$ is assumed to be on the far side of P_2 on the P_1–P_2 line. Breaking up a four-body problem into two disjoint two-body problems is an enormous simplification to the capture problem and dynamically is not correct. Nevertheless, these transfers change little when they are applied with full solar system modeling in many useful cases. This is because of the high energy associated with them. Their derivation is elegantly simple, and their usefulness is remarkable. They have paved the way for both human and robotic exploration of our solar system.

Of particular interest for applications considered later in this book is when r_{13}, r_{23} are relatively small numbers. Let km = kilometer, s = second. For example, $r_{13} = r_E + 200$ km, $r_{23} = r_M + 100$ km are typical radial distances used in applications of P_3 from P_1, P_2 at the locations $\mathbf{Q_0}$, $\mathbf{Q_F}$, respectively, and at times $t = t_0, t_F$, respectively. r_E, r_M represent the radii of the Earth, Moon respectively. As is verified in this case from [118], $\Delta V_0 = 3.142$ km/s, $\Delta V_1 = 0$. $\Delta V_C = .200$ km/s, .648 km/s for $e_2 = .95, 0$, respectively. Also, $t_F - t_0 = 5$ days. The transfer itself is nearly parabolic where $e_1 = .97$. Visually it would appear to be nearly linear. For Hohmann transfers in general, $E_2 \gg 0$ at $\mathbf{Q_F}$. These values of r_{13}, r_{23} are the values that we desire for a solution of the capture problem.

It is verified that $E_2 > 0$ at $\mathbf{Q}_F$ for P_3, and this causes a large value of ΔV_C to occur. This property of $E_2 > 0$ is satisfied by Hohmann transfers. The reason $E_2 > 0$ follows from the fact that the magnitude V_F of the velocity vector at $\mathbf{Q_F}$ of P_3 on the transfer at lunar periapsis, where the direction is in the same direction as the Moons orbit about the Earth, has the property that $V_F \ll V_M$, where V_M is the magnitude of the velocity of the Moon about the Earth. It turns out that under the given assumptions, $V_F = 0.176$ km/s and $V_M = 1.019$ km/s. This implies that $E_2 = .843$ km^2/s^2. It is the discrepancy between V_F and V_M that yields a large value of ΔV_C of several hundred meters per second, depending on the value of e_2. The calculation of E_2 for a Hohmann transfer is estimated by noting that relative to P_2, the transfer is hyperbolic, with a hyperbolic periapsis at $\mathbf{Q_F}$. The corresponding velocity at $r_2 = \infty$, called the hyperbolic excess velocity and labeled V_∞, is estimated by $V_\infty = V_M - V_F = .843$ km/s yielding $E_2 = (1/2)V_\infty^2$. The calculation of ΔV_C follows from a functional relationship it has with V_∞, or equivalently E_2 (See [20]).

A ballistic capture transfer will arrive at periapsis at P_2 where E_2 is negative, and therefore it will have no V_∞. This enables capture where $\Delta V_C = 0$. Eliminating the V_∞ is the motivation for the construction of ballistic capture transfers. We will see that a ballistic capture transfer going from $\mathbf{Q}_0$ to $\mathbf{Q}_F$ can be constructed with approximately the same value of ΔV_0 for a Hohmann transfer, and with $\Delta V_1 = 0$. It has the key property that $E_2 < 0$, where $e_2 \approx 0.95$. This turns out to have important implications for the construction of transfers in astrodynamics.

Prior to 1986, the construction of practical ballistic capture transfers was thought to be impossible; in that year one was numerically constructed that is described in section 3.3.2. The methodology of its construction led to the design of a more interesting transfer that was actually used in 1991, which is described in section 3.4.

Note that we refer to a Hohmann transfer as *high energy* since V_∞ is significantly high, and a ballistic capture transfer is called *low energy* since the V_∞ is eliminated. This is a fundamental difference between these two types of transfers.

3.2 THE WEAK STABILITY BOUNDARY

Ballistic capture is defined in section 3.1 for the elliptic restricted problem. This type of capture occurs at the weak stability boundary W defined for the planar circular restricted problem (1.61) by (3.9). We now restrict ourselves to the planar circular restricted problem.

W is defined by (3.9) in barycentric rotating coordinates (1.59). For ease of calculations, we will derive the expression for W in P_2-centered rotating coordinates. For W to be well defined, the Jacobi constant C needs to be restricted. The range of C is estimated where W exists. This yields an estimate of the range of C where ballistic capture occurs.

3.2.1 Numerical Algorithmic Definition

To motivate an analytic approximation of W in a P_2-centered rotating coordinate system in the planar circular restricted problem, we give a definition based upon numerical exploration [25, 35].

Consider a radial line ℓ from P_2 as shown in Figure 3.5 in a P_2-centered rotating coordinate system X_1, X_2. We follow trajectories $\varphi(t)$ of P_3 starting on ℓ, which satisfy the following requirements.

- The initial velocity vector of the trajectory for P_3 is normal to the line ℓ, pointing in the direct (posigrade) or retrograde directions.
- The initial two-body Kepler energy E_2 of P_3 with respect to P_2 is negative or 0, where E_2 is given by (3.6).
- The eccentricity $e_2 \in [0, 1]$ of the initial two-body Keplerian motion is fixed along ℓ. The initial velocity magnitude $V_2 = (\dot{X}_1{}^2 + \dot{X}_2{}^2)^{\frac{1}{2}} = (\mu(1+e_2)/r_{23})^{\frac{1}{2}} - r_{23}$, $0 < \mu \leq 1/2$. It varies along ℓ. The term r_{23} is subtracted from the inertial velocity since this is a rotating system.

Thus P_3 starts its motion on an osculating ellipse which we assume is at its periapsis. Hence,

$$E_2 = \frac{\mu}{2}\left(\frac{e_2 - 1}{r_{23}}\right) \leq 0. \tag{3.12}$$

This value of E_2 is the general value for an elliptic orbit obtained from (1.29), with $k = \mu$, and $r_{23} = a(1 - e_2)$ is the periapsis. For notational convenience, we set $r_{23} = r_2$. We define stability in the following way. Suppose that the initial position of P_3 is at the point $a \in \ell$ in Figure 3.5 at initial time $t = t_0$.

The motion of P_3 is *stable* about P_2 if

(i) after leaving ℓ it makes a full cycle about P_2 without going around P_1 and returns to a point $b \in \ell$, where $E_2 \leq 0$.

The motion of P_3 is *unstable* if either

(ii) it performs a full cycle about P_2 without going about P_1 (i.e., $\theta_1 \neq 0$, where θ_1 is the polar angle with respect to P_1 shown in Figure 3.5) and returns to a point $b \in \ell$, where $E_2 > 0$; or

(iii) P_3 moves away from P_2 towards P_1 and makes a cycle about P_1 achieving $\theta_1 = 0$, or P_3 collides with P_1. It is assumed that for $t > t_0$, once P_3 leaves ℓ, where $\theta_2 = \theta_2(t_0) \in [0, 2\pi)$, P_3 need only cycle about P_2 until $\theta_2(t) = 2\pi$.

It is noted that (i) corresponds to ballistic capture at b with respect to P_2 by Definition 3.11, and the orbit from a to b is a ballistic capture transfer by Definition 3.13, which by Definition 3.15 is bounded. (ii) corresponds to ballistic escape from P_2. (iii) represents a different type of escape.

Definition 3.19 *The escape in case (iii) is called* primary interchange escape.

Primary interchange escape need not imply ballistic escape since it may be the case that $E_2(\varphi(t))$ remains negative or 0 up to the time that $\theta_1 = 0$. The case where P_3 collides with P_1 is a degenerate case of primary interchange escape. By regularization this case is well defined.

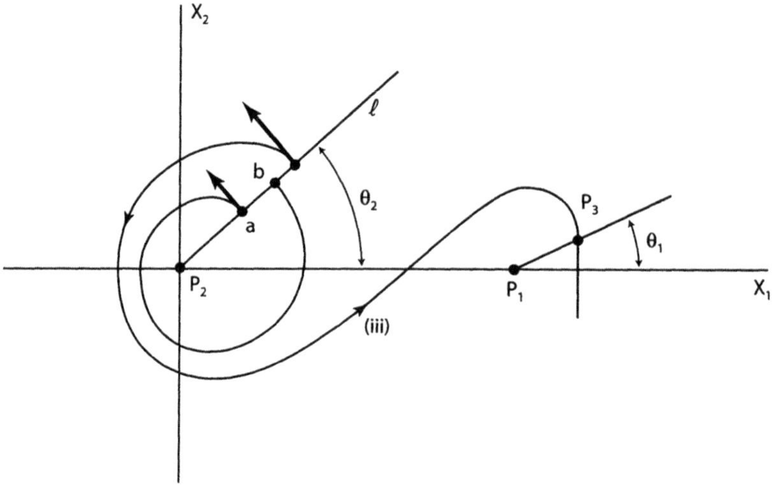

Figure 3.5 Stable, unstable motion and primary interchange escape.

As the initial conditions vary along ℓ satisfying (i), (ii), (iii), it is numerically found that there is a finite distance r^* on ℓ from P_2 satisfying the following statements:

If $r_2 < r^*$, the motion is stable.

If $r_2 > r^*$, the motion is unstable.

r^* depends on only two parameters, the polar angle θ_2 which ℓ makes with the x_1-axis and the eccentricity e_2 of the osculating Keplerian ellipse at the point a at $t = t_0$. r_2 is determined to be a well-defined function of θ_2, e_2. Define

$$\mathcal{W} = \{r^*(\theta_2, e_2) \in \mathbf{R}^1 | \theta_2 \in [0, 2\pi], e_2 \in [0, 1]\}. \quad (3.13)$$

$\mathcal{W}$ is a two-dimensional stability transition region of position and velocity space, which we call the *weak stability boundary.* We have not shown dependence on μ since μ is fixed. $\mathcal{W}$ has two components. One corresponds to retrograde motion about P_2 and the other to direct motion about P_2 after propagation from ℓ. $\mathcal{W}$ is described in [35].

It is noted that if we had considered the two-body problem between P_3 and P_2 ignoring P_1, then $r^* = \infty$. Thus, the effect of the gravitational perturbation of P_1 is to allow escape to occur with respect to P_2 at a finite distance.

It is also noted that E_2 in (3.6) is defined in P_2-centered inertial coordinates X_1, X_2, not to be confused with the rotating coordinates used in the definition of the algorithm. It is convenient to transform E_2 to P_2-centered rotating coordinate system $\bar{x}_1, \bar{x}_2$. This is done by applying (1.59) and replacing $\mathbf{x}$ by $\bar{\mathbf{x}}$, and $\mathbf{Q}$ by $\mathbf{X}$, obtaining

$$\hat{E}_2(\bar{\mathbf{x}}, \dot{\bar{\mathbf{x}}}) = E_2\left(R^{-1}\bar{\mathbf{x}}, \frac{d}{dt}(R^{-1}\bar{\mathbf{x}})\right), \tag{3.14}$$

or more specifically

$$\hat{E}_2(\bar{\mathbf{x}}, \dot{\bar{\mathbf{x}}}) = \frac{1}{2}|\dot{\bar{\mathbf{x}}}|^2 - \frac{\mu}{r_2} + \frac{1}{2}r_{23}^2 - L, \tag{3.15}$$

where $L = \dot{\bar{x}}_1\bar{x}_2 - \dot{\bar{x}}_2\bar{x}_1$ and $r_2 = |\bar{\mathbf{x}}|$. The transformation $\bar{\mathbf{x}} = \mathbf{T}(\mathbf{x})$ to barycentric rotating coordinates x_1, x_2 in (1.59) is just the translation

$$\bar{x}_1 = x_1 - (-1 + \mu), \qquad \bar{x}_2 = x_2, \tag{3.16}$$

where $\tilde{E}_2(\mathbf{x}, \dot{\mathbf{x}}) = \hat{E}_2(\mathbf{T}(\mathbf{x}), \dot{\mathbf{x}})$. Recall that $\hat{E}_2$ is given as $\tilde{H}$ in (1.71) with $a \equiv \mu$.

Definition 3.19 is extended in a more general way to the case of capture. In Figure 3.5, the coordinate system is rotating so that P_1, P_2 are fixed. Let $\phi(t)$ denote the trajectory of P_3 in phase space. We refer to Figure 3.5; the line ℓ is not considered.

Definition 3.20 *At* $t = t_1$ *assume* $\theta_1 = 0$, $r_{13} > 0$, *and at* $t = t_2 > t_1$, $\theta_1 \geq \pi$, $E_2(\phi(t_2)) \leq 0$. *Then* P_3 *has a* primary interchange capture *at* $t = t_2$. *The trajectory* $\phi(t)$ *for* $t \in [t_1, t_2]$ *is called a* primary interchange capture transfer.

Thus, a primary interchange transfer requires *both* ballistic capture at $t = t_2$ at P_2 and also that the trajectory of P_3 in physical space achieves a partial cycling about P_1 of at least 180 degrees.

3.2.2 Analytic Approximation W

In barycentric rotating coordinates, $\tilde{J}(\mathbf{x},\dot{\mathbf{x}})$ is given by (1.63). Using (3.16) $\tilde{J}$ is transformed into $\bar{J}(\bar{\mathbf{x}},\dot{\bar{\mathbf{x}}}) = \tilde{J}(\mathbf{T}^{-1}(\bar{\mathbf{x}}),\dot{\bar{\mathbf{x}}})$. It is verified that

$$\bar{J} = -2\left[\frac{1}{2}|\dot{\bar{\mathbf{x}}}|^2 - \frac{\mu}{|\bar{\mathbf{x}}|}\right] + \bar{A}(\bar{x},\bar{y}), \tag{3.17}$$

$$\bar{A}(\bar{x},\bar{y}) = (\bar{x}_1 - 1 + \mu)^2 + \bar{x}_2^2 + 2\frac{1-\mu}{r_{13}}, \tag{3.18}$$

where $r_{13}^2 = (x_1 - 1)^2 + x_2^2$. From (3.14), (3.17) can be written as a function of $\hat{E}_2$,

$$\bar{J} = -2\left[\hat{E}_2 + L - \frac{1}{2}|\bar{\mathbf{x}}|^2\right] + \bar{A}. \tag{3.19}$$

Let $r_2 = |\bar{\mathbf{x}}|$ represent the radial distance of P_3 to the origin where P_2 is located, θ_2 the polar angle with respect to the $\bar{x}_1$-axis,

$$\bar{x}_1 = r_2\cos\theta_2, \qquad \bar{x}_2 = r_2\sin\theta_2, \tag{3.20}$$

$\theta_2 \in [0, 2\pi], r_2 \geq 0$, and e_2 is the eccentricity of the orbit of P_3 with respect to P_2. Equation (3.19) is expressed in polar coordinates,

$$\mathcal{J} = -2\left[E_2^* + L^* - \frac{1}{2}r_2^2\right] + A, \tag{3.21}$$

where E_2^*, L^*, A are $\hat{E}_2, L, \bar{A}$, respectively, expressed in polar coordinates (3.20):

$$\begin{aligned}
E_2^*(r_2,\theta_2,\dot{r}_2,\dot{\theta}_2) &= \frac{1}{2}(\dot{r}_2^2 + r_2^2\dot{\theta}_2^2) - \frac{\mu}{r_2},\\
L^*(r_2,\dot{\theta}_2) &= -r_2^2\dot{\theta}_2,\\
A(r_2,\theta_2) &= (r_2\cos\theta_2 - 1 + \mu)^2 - r_2^2\sin^2\theta_2 + 2\frac{1-\mu}{r_{13}},
\end{aligned}$$

where $r_{13} = (r_2\cos\theta_2 - 1)^2 + r_2^2\sin^2\theta_2$. A is periodic of period 2π in θ_2.

We now restrict the variables $r_2, \dot{r}_2, \theta_2, \dot{\theta}_2$ to special sets to be consistent with the numerical algorithm. The first set is

$$\Sigma^* = \left\{(r_2,\theta_2,\dot{r}_2,\dot{\theta}_2) \in B | E_2^* = \frac{\mu}{2}\left(\frac{e_2 - 1}{r_2}\right) \leq 0\right\}, \tag{3.22}$$

where

$$B = \left\{r_2 \geq 0, \theta_2 \in [0,2\pi], \dot{r}_2 \in \mathbf{R^1}, \dot{\theta}_2 \in \mathbf{R^1}, e_2 \in [0,1], 0 < \mu < 1/2\right\}.$$

Σ^* is defined since we propagated solutions from the line ℓ where the two-body energy E_2 satisfies the condition (3.12). The Jacobi integral surface is

$$\mathcal{J}^{-1}(C) = \left\{(r_2,\theta_2,\dot{r}_2,\dot{\theta}_2) \in B | \mathcal{J} = C \in \mathbf{R^1}\right\}. \tag{3.23}$$

The set $\mathcal{J}^{-1}(C) \cap \Sigma^*$ yields points on $\mathcal{J}^{-1}(C)$ where $E_2^* \leq 0$. Thus

$$\mathcal{J}|_{\Sigma^*} = -2\left[\frac{\mu}{2}\left(\frac{e_2-1}{r_2}\right) - r_2^2\dot{\theta}_2 - \frac{1}{2}r_2^2\right] + A. \tag{3.24}$$

Points are considered where the elliptic states are at periapsis, $\dot{r}_2 = 0$, in accordance with the numerical algorithm. This means that the magnitude of the velocity relative to P_2 is $V_2 = (\dot{r}_2^2 + r_2^2\dot{\theta}_2^2)^{\frac{1}{2}} = r_2\dot{\theta}_2$, and since P_3 is at a periapsis,

$$r_2\dot{\theta}_2 = \pm\sqrt{\frac{\mu(1+e_2)}{r_2}} - r_2, \tag{3.25}$$

where r_2 is subtracted since we are in a rotating coordinate system now centered at P_2, and $\pm$ correspond to direct, retrograde motion, respectively.

Since the frequency of the rotation is $\omega = 1$, then at a distance r_2 from the origin which is at a periapsis state, the velocity vector is normal to a radial line, and the magnitude of the rotational velocity is $r_2\omega = r_2$, which is subtracted from the inertial velocity magnitude. Thus, restricting the points of the phase space to the set

$$\sigma^* = \left\{(r_2, \theta_2, \dot{r}_2, \dot{\theta}_2) \in B | \dot{r}_2 = 0\right\} \tag{3.26}$$

yields (3.25). Equation (3.25) eliminates $\dot{\theta}_2$ from (3.24) and can be viewed as a transformation of

$$\dot{\theta}_2 \to e_2$$

for each fixed r_2. Thus, it is verified that

$$\mathcal{J}|_{\Sigma^* \cap \sigma^*} = -r_2\left[\pm 2\sqrt{\frac{\mu(1+e)}{r_2}} + r_2\right] + \mu\frac{1-e}{r_2} + A. \tag{3.27}$$

This represents a transformation of $\mathcal{J}$ in the coordinates $r_2, \theta_2, \dot{r}_2, \dot{\theta}_2 \to r_2, \theta_2, e_2$.

Hence, for each fixed μ, C, the set

$$W = \mathcal{J}^{-1}(C) \cap \Sigma^* \cap \sigma^* \tag{3.28}$$

is two-dimensional in the space (r_2, θ_2, e_2) and is given by the expression

$$C = -r_2\left[\pm 2\sqrt{\frac{\mu(1+e_2)}{r_2}} + r_2\right] + \mu\frac{1-e_2}{r_2} + A(r_2, \theta_2), \tag{3.29}$$

where $\theta_2 \in [0, 2\pi], e_2 \in [0, 1], r_2 \geq 0, 0 < \mu < 1/2$. We have proven the following lemma.

Lemma 3.21 *The Jacobi integral (3.21) restricted to the set (3.28) is expressed by (3.29) which is an explicit expression for* W.

The set $W = \mathcal{J}^{-1}(C) \cap \Sigma^* \cap \sigma^*$ represents an approximation of the weak stability boundary $\mathcal{W}$ provided the values of C are suitably restricted.

The necessity of restricting C follows from consideration of the *Hill's regions* of the restricted problem. Let's return to the differential equations of the restricted problem in barycentric rotating coordinates x_1, x_2 from (1.61). These differential equations have five equilibrium points where the vector field vanishes when (1.61) is written as a first order system of four differential equations, $\dot{x}_k = f_k(\mathbf{x}), k = 1,2,3,4, \mathbf{x} = (x_1, x_2, x_3, x_4), \dot{x}_1 = x_3, \dot{x}_2 = x_4$, and $\dot{x}_3 = f_1, \dot{x}_4 = f_2$ are given by (1.61). That is, they are found by solving the system $f_k(\mathbf{x}) = 0, k = 1,2,3,4$. Also, it is recalled from section 1.5 that the right-hand side of (1.61) is the sum of the centrifugal force $\mathbf{F}$ and the sum of the gravitational forces $\mathbf{G}$. The five equilibrium points can be viewed as being five locations where P_3 is at rest and $\mathbf{F}+\mathbf{G} = \mathbf{0}$. These points are well known in the literature, where they are labeled $L_k, k = 1,2,3,4,5$ and called Lagrange points [219]. Three are collinear, $L_k, k = 1,2,3$, and two, L_4, L_5, lie at the vertices of equilateral triangles. They are shown in Figure 1.9. They were collectively discovered by Lagrange and Euler [132, 80]. Note that Euler's work was published in the 16th century, while Lagrange's work was published over 100 years later, in the 17th century.

The equilibrium points $L_k, k = 1,2,3$, are locally unstable and are saddle-center points. L_4, L_5 are locally stable. The proof of the instability of L_1, L_2, L_3 is due to Conley [64]. The proof of stability of L_4, L_5 is due to Deprit and Deprit-Bartolomé [67].

The points of interest for our considerations are L_1, L_2. Evaluating $\tilde{J}$ at L_k where $\dot{\mathbf{x}} = 0$ yields the corresponding values $\tilde{J} = C_k, k = 1,2,3,4,5$, satisfying the relative ordering

$$C_4 = C_5 = 3 < C_3 < C_1 < C_2, \tag{3.30}$$

where we are considering the integral $\tilde{J} \equiv \tilde{J} + c_0, c_0 = \mu(1-\mu)$ with the additive constant c_0 in order to uniformize the values of C_k to be consistent with those in [219]. *This is assumed until further notice.* This additive constant has no effect on the form of the differential equations.

In fact, an approximation for C_1 and C_2 valid to three digits when $\mu \leq 0.01$ or four digits when $\mu \leq 0.001$ is

$$C_1 \approx 3 + 9\left(\frac{\mu}{3}\right)^{\frac{2}{3}} - 11\left(\frac{\mu}{3}\right), \quad C_2 \approx 3 + 9\left(\frac{\mu}{3}\right)^{\frac{2}{3}} - 7\left(\frac{\mu}{3}\right). \tag{3.31}$$

Equation (3.31) implies for $0 < \mu \ll 1, C_1, C_2 \gtrsim 3$. C_k play a key role in describing the regions of motion in x_1, x_2-space where P_3 can move, called the

Hill's regions. The manifold $\tilde{J}^{-1}(C)$ projected onto the x_1, x_2-plane form the Hill's regions

$$\mathcal{H}(C) = \left\{ \mathbf{x} \in \mathbf{R^2} | 2\hat{\Omega} - C \geq 0 \right\},$$

where

$$\hat{\Omega} = \Omega + \frac{1}{2}|\mathbf{x}|^2 + \frac{1}{2}c_0.$$

The inequality in $\mathcal{H}$ results from the fact that $\tilde{J} = -|\dot{\mathbf{x}}|^2 + 2\hat{\Omega}$, and $2\hat{\Omega} - C = |\dot{\mathbf{x}}|^2 \geq 0$. P_3 is constrained to move within $\mathcal{H}(C)$. The boundary of $\mathcal{H}(C) = \partial\mathcal{H}$ is given by the curves

$$Z(C) = \left\{ \mathbf{x} \in \mathbf{R^2} | 2\hat{\Omega} - C = 0 \right\},$$

which are called *zero-velocity curves* since $|\dot{\mathbf{x}}| = 0$.

An excellent description of the Hill's regions and how they evolve as C varies is given in [219]. We give a very brief description here. The qualitative appearance of $\mathcal{H}$ as C varies is shown in Figures 3.6 (a)–(e). In these figures P_3 cannot move in the shaded regions. Thus, as C varies, one obtains domains where P is constrained to move.

For example, for $C > C_2$, P_3 cannot pass between P_1, P_2. This is shown in (a). P_3 is constrained to move in a neighborhood of P_2 or P_1, or to move around both the mass points beyond the outer circular-type curve. When $C \lesssim C_2$ P_3 can pass from P_1 to P_2. This is shown in (b). This case is of interest in section 3.3. C_2 represents the upper bound of the minimal energy such that primary interchange capture can occur from P_1 to P_2 for $C \lesssim C_2$, where P_2 can pass through a small opening near L_2. P_3 still can move about both the primary masses as well, in the *outer region*; however, it cannot pass from the outer region to either P_1 or P_2. The critical value of C for this to happen is when $C = C_1 < C_2$. When $C \lesssim C_1$, P_3 can pass from the peanut-shaped region connecting P_1, P_2 with the outer region through the opening near L_1. This is seen in (c). For $C < C_1$ unbounded escape and capture can therefore occur between P_3 and either of the two primaries. This case is of relevance to section 3.4. For $C \geq C_1$, $\mathcal{H}$ has two components, one containing bounded motion and the other allowing unbounded motion. For $C < C_1$ $\mathcal{H}$ has one component. Part (d) corresponds to when $3 \leq C < C_1$. In (e) $C < 3$ and the Hill's region is the entire plane, and P_3 can move anywhere.

Thus, we have by Lemma 3.21 the following definition.

Definition 3.22 *The set* $W = \mathcal{J}^{-1}(C) \cap \Sigma^* \cap \sigma^*$ *which is given by (3.29) represents an approximation of the weak stability boundary* $\mathcal{W}$ *provided* $C < C_1$.

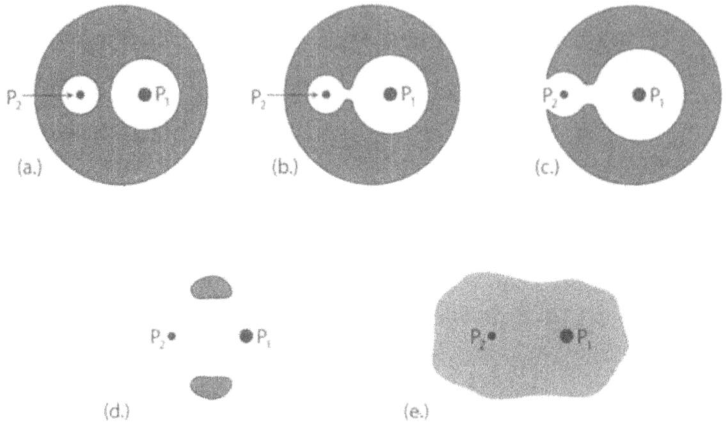

Figure 3.6 Hill's regions.

Numerical evidence indicates that W represents a good approximation of $\mathcal{W}$ [32, 31]. It is remarked that as opposed to the Lagrange points where $\mathbf{F}+\mathbf{G}=\mathbf{0}$ and $\dot{\mathbf{x}}=0$, on or near $\mathcal{W}$, $\mathbf{F}+\mathbf{G}\approx 0$ and $\dot{\mathbf{x}}\neq\mathbf{0}$. This shows how $\mathcal{W}$ represents a generalization of the Lagrange points.

Definition 3.22 does not imply that W actually exists since it is necessary that $C < C_1$, and it may be W is the empty set. This defines the *C-problem.*

Equation (3.29) implies

$$C = C(r_2, \theta_2, e_2), \tag{3.32}$$

where C is periodic of period 2π in θ_2. The dependence on μ is not shown since it is fixed. Let $\mathbf{R}^+$ be the set of real numbers $x \geq 0$.

Definition 3.23 *The C-problem is given by the existence of the three-dimensional manifold,*

$$Q_1 = \left\{(r_2, \theta_2, e_2) \in \mathbf{R}^+ \times [0, 2\pi] \times [0, 1] | C(r_2, \theta_2, e_2) < C_1\right\}.$$

Q_1 is shown to exist by graphing (3.32), as will be seen, where C has a finite lower bound depending on μ.

Lemma 3.24 *Equation (3.29) implicitly yields*

$$r_2 = r_2(\theta_2, e_2, C), \tag{3.33}$$

where r_2 is periodic of period 2π in θ_2, and $(r_2, \theta_2, e_2) \in Q_1$.

Equation (3.33) implies

Lemma 3.25 *For each value of $C < C_1$, $W \subset \mathcal{J}^{-1}(C)$ is generally two-dimensional and is the union of two-dimensional topological annuli or discs parameterized by $(\theta_2, e_2), \theta_2 \in [0, 2\pi], e_2 \in [0, 1]$. These annuli or discs are defined by solutions of (3.33) and are centered at P_2.*

It is remarked that W can have one-dimensional components in degenerate cases which we omit.

By fixing C and letting r_2, θ_2, e_2 vary in the set Q_1, it is verified that an explicit embedding of W into four-dimensional phase space $\mathbf{R}^4$ with coordinates $x_1, x_2, \dot{x}_1, \dot{x}_2$ can be constructed. The embedding is given by a map $x_i = x_i(r_2, e_2, \theta_2), \dot{x}_i = \dot{x}_i(r_2, e_2, \theta_2)$ of $Q_1 \to \mathbf{R}^4$.

We can view W as a location where P_3 has elliptic or parabolic initial conditions with respect to P_2, and as time increases P_3 may ballistically escape P_2 or do a primary interchange. The boundary of W in C-space is $C = C_1$.

For $r_2 \gtrsim 0$ it is seen that $A \approx 3 - 3\mu$ by (3.18) since $r_{13} \approx 1, (\bar{x}_1 - 1 + \mu)^2 + \bar{x}_2^2 \approx (1-\mu)^2$, where we have noted that $\bar{J}$ now has the additive term $c_0 = \mu(1-\mu)$.

Lemma 3.26 *For $r_2 \gtrsim 0, C \lesssim C_1, \mu \ll 1$, then*

$$r_2 \approx \frac{(1-e_2)\mu^{\frac{1}{3}}}{3^{\frac{5}{3}} - \frac{2}{3}\mu^{\frac{1}{3}}}. \tag{3.34}$$

Proof. Substitute $A \approx 3 - 3\mu$ into (3.29),

$$C \approx -r_2 \left[\pm 2\sqrt{\frac{\mu(1+e_2)}{r_2}} + r_2 \right] + \mu \frac{1-e_2}{r_2} + 3 - 3\mu.$$

If $C \lesssim C_1 \Rightarrow C \approx C_1$. Substitute the expression for C_1 in (3.31) into the previous equation, which implies

$$\begin{aligned} 3 + 9\left(\frac{\mu}{3}\right)^{\frac{2}{3}} - 11\left(\frac{\mu}{3}\right) + \cdots \\ \approx -r_2 \left[\pm 2\sqrt{\frac{\mu(1+e)}{r_2}} + r_2 \right] + \\ \mu\frac{1-e_2}{r_2} + 3 - 3\mu. \end{aligned}$$

For $\mu \ll 1, r_2 \gtrsim 0$ the first term on the right hand side is ignored since it is of higher order. Ignoring the higher order terms on the left hand side and solving for r_2 yields the proof of the Lemma. □

3.2.3 Visualization of W

We graph W using (3.29) for the Earth–Moon system where $\mu = 0.0123$. The easiest way to do this is to fix θ_2 and to let (r_2, e_2) vary. A representative one is shown in Figure 3.7, where $\theta_2 = \pi$. The points plotted are $(e_2, r_2, -C)$, where $-C \in [-C_1, -C_1 + 0.4], r_2 \in [0, .99]$, and $e_2 \in [0, 1]$. Note that the Lagrange points L_i are at a distance d_i from P_2 for $i = 1, 2$, where $d_1 = \alpha^{1/3}[1 + \frac{1}{3}\alpha^{1/3} + \cdots], d_2 = \alpha^{1/3}[1 - \frac{1}{3}\alpha^{1/3} + \cdots]$ and $\alpha = \mu/3$. Since $\mu = 0.0123$, we see that $d_i \approx 0.169$.

In Figure 3.7 the value of $-C$ along the vertical axis starts at $-C_1$. For the Earth–Moon system $C_1 \approx 3.184$. We solve the C-problem by determining where the values of $-C(r_2, \theta, e_2)$ are greater than $-C_1$. This means that the W surface lies above the (r_2, e_2)-plane, which happens for a substantial range of r_2 depending on (e_2, θ_2). For fixed values of e_2 and θ_2, r_2 lies in finite intervals. When the graph is flat and lies on the (r_2, e_2)-plane, then W lies below the plane $\{C = -C_1\}$. This is plotted using *Mathematica*.

In Figure 3.7 we see that as e increases to 1, the lower limit of r_2 decreases towards zero. This is exactly what is observed numerically [35]. Moreover, the curve $r_2(e_2)$ for the lower limit of r_2 is very close to numerical results and, more importantly, to the applied results we consider in later sections. The W surface generally has two basic types of appearance as θ_2 varies. One type is that which is shown for $\theta_2 = \pi$. This changes as θ_2 varies about $\frac{\pi}{2}, \frac{3\pi}{2}$. In these cases, instead of the surface being parabolic in shape, increasing near the e_2-axis where it intersects the plane $C = C_1$, rising to a maximum, then decreasing below the $C = C_1$ plane, it keeps on increasing until $r_2 = 1$ is reached.

When $\theta_2 = \frac{\pi}{2}, \frac{3\pi}{2}$ the W surface has a C range of maximal size, where $C \in [2.22, C_1]$, occuring for $e_2 = 1$. The W surface is the same for $\theta_2 = \frac{\pi}{2}$ and $\frac{3\pi}{2}$ as is verified.

The range of the parameters for two cases are listed here for the two boundary values of $e_2 = 0, 1$ rounded off to the hundredths place. One case is for $\theta_2 = \pi$, which we list since it is of importance later in the book. The other is for $\theta_2 = \frac{\pi}{2}$ since it yields the maximal C variation. The C ranges for any value of θ_2 take on a maximal variation for $e_2 = 1$ and a minimal variation for $e_2 = 0$.

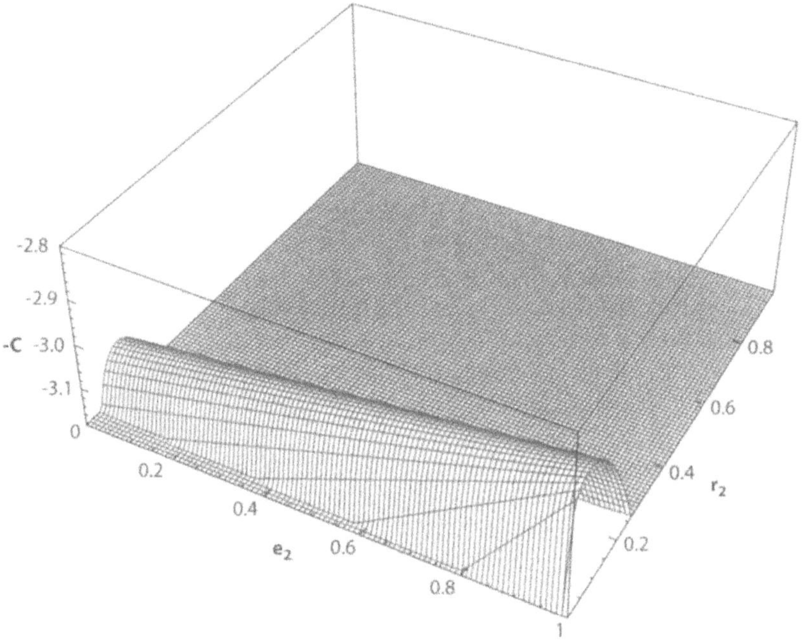

Figure 3.7 Moon W surface: $(e_2, r_2, -C), \theta_2 = \pi$.

For $\theta = \pi$, on the $e_2 = 0$ axis we have $C \in [2.98, C_1]$ and $r_2 \in [0.04, 0.50]$, while on the $e_2 = 1$ axis we have $C \in [2.88, C_1]$ and $r_2 \in [0.00, 0.58]$.

For $\theta = \frac{\pi}{2}$, on the $e_2 = 0$ axis, we have $C \in [2.25, C_1]$ and $r_2 \in [0.05, 1.00]$, while on the $e_2 = 1$ axis we have $C \in [2.22, C_1]$ and $r_2 \in [0.00, 1.00]$.

Note that C need not be only slightly less than C_1 but can decrease to approximately 2.22. Thus, the W surface exists for $C \in [2.22, C_1]$ representing a solution to the C-problem. Analogous results are obtained for different values of μ. Note that for $C < 3$ the zero-velocity curves disappear and P_3 is free to move in the entire x_1, x_2-plane.

3.3 EXISTENCE OF PRIMARY INTERCHANGE CAPTURE AND AN APPLICATION

In this section we show how to theoretically prove the existence of low energy lunar transfers from P_1 (Earth) to P_2 (Moon) for the restricted three-body problem. These transfers have the unique property that upon arrival at a point $\mathbf{Q}_F$ near $P_2, E_2 < 0$ so that the V_∞ value of the Hohmann transfer is eliminated. These transfers represent a substantially less energetic solution to the capture problem than the Hohmann transfer.

A presentation of how to achieve a low energy transfer from the Earth to the Moon for the planar circular restricted problem is given by Conley [63]. He does not describe this in terms of ballistic capture, and his analysis is local only. He uses the existence of invariant manifolds near the Lagrange point L_2 for $C \lesssim C_2$. This is described in subsection 3.3.1. We prove that local ballistic capture follows directly from his formulation. This existence of local ballistic capture is extended to a proof of existence of primary interchange capture which is global in nature, extending Conley's results. More general results by Llibre, Simó, and Martinez are discussed.

Conley conjectures how numerical investigations may produce transfer orbits which could be useful for applications by passing from near P_1 to near L_2. This is not carried out in [63]. Such applications are also discussed in more depth by Heppenheimer [98], who numerically investigated orbits that pass between Hill's region about P_1, P_2 for $C \lesssim C_2$ in the planar restricted problem. However, this is not carried out for realistic conditions, and a precision case is not constructed. The notion of ballistic capture is also not considered in [98].

A precision numerically generated example of a ballistic capture transfer starting near P_1 and passing near L_2 close to P_2 in the Hill's region about P_2 for $C \lesssim C_2$ is constructed in [25] and described here. This represents the first known example of a realistic ballistic capture transfer. It is more precisely a three-dimensional primary interchange capture transfer. In addition, it goes to ballistic capture at the weak stability boundary. The modeling is realistic and uses a planetary ephemeris [212] to accurately model the motion of the Moon, Earth, and P_3 in three dimensions. This is described in subsection 3.3.2. This example provides a suitable solution to the questions Conley addressed on applications.

The ballistic capture transfer used in this application has a starting point approximately $r_{13} = 150,000$ km from the Earth so that it *does not* represent a desired solution to the capture problem where we desire $r_{13} = r_E + 200$ km. This latter value of r_{13} is the type which is useful for applications to go to the Moon and beyond.

The desired ballistic capture transfers with starting values $r_{13} = r_E + 200$ km are constructed in section 3.4 and require a four-body problem for the model.

3.3.1 Proof of Existence of Primary Interchange Capture Transfers

Conley wrote an interesting paper on the stability of the collinear Lagrange points [64]. Prior to Conley's work it was well known that the Lagrange points L_1, L_2, L_3 are locally linearly unstable, being locally linearly saddle-centers, and L_4, L_5 are stable being locally linearly elliptic [136, 219]. In the planar circular restricted problem that we consider, the fact that L_1, L_2, L_3 are locally saddle-centers means that if the differential equations of the restricted problem (1.61) are linearized about these equilibrium points, the eigenvectors have nonzero real and imaginary parts. They are therefore of the form $\pm\lambda, \pm i\nu, \lambda, \nu \in \mathbf{R}^1$. The real parts give rise to the unstable motion, and the imaginary parts give rise to the stable motion [14]. In the case of L_4, L_5 the eigenvalues are purely imaginary.

Local linearized behavior of a vector field near a saddle equilibrium point generally implies the same behavior for the full nonlinear vector field as follows from the stable manifold theorem. However L_1, L_2, L_3 are saddle-centers. Nevertheless Conley proved that this structure is preserved for the full vector field. This follows by a theorem of Moser [171] on a generalization of a theorem of Lyapunov. This is called the Moser-Lyapunov theorem and it is applicable to a general Hamiltonian system of two degrees of freedom with an equilibrium point. Conley proved that the Moser-Lyapunov theorem was applicable to L_1, L_2, L_3 and showed that the Hamiltonian function of the circular restricted problem when put in linearized form about $L_k, k = 1, 2, 3$, has the required form for the Moser-Lyapunov theorem. This is formulated in the following theorem.

Theorem 3.27 (Conley) *The flow of the restricted problem in a neighborhood of $L_k, k = 1, 2, 3$, is equivalent to the linearized flow. This is true for $C = C_k$ and $C \lesssim C_k$.*

Of particular interest is the flow of the restricted problem (1.61) in a neighborhood of L_2 for $C \lesssim C_2$.

From examination of the Hill's regions in Figure 3.6, it is seen that for $C \lesssim C_2$, the Hill's region about P_1, P_2 becomes connected near L_2, creating what Conley terms a *neck*. For $C \lesssim C_2, P_3$ can then pass through the neck from the part of the Hill's region about the Earth, which we call $\mathcal{H}_E$, to the part about the Moon, $\mathcal{H}_M$, and vice versa. With this energy it is possible to construct trajectories passing between the Earth and Moon.

When $C \lesssim C_2$, L_2 disappears, and bifurcating from it is a family $\mathcal{L}_2$ of retrograde periodic orbits parameterized by the energy $C, \mathcal{L}_2 = \mathcal{L}_2(C)$

[219, 136]. This family also depends on $\mu \in (0, 1/2)$ but this variable is not shown since it is fixed. These are called the *family of Lyapunov orbits*. These are proven to exist in the linearized problem [219] and therefore by Theorem 3.27 they exist in the full restricted problem. They are unstable due to the real eigenvalues and have two-dimensional stable and unstable manifolds $\mathbf{W^s}(\mathcal{L}_2), \mathbf{W^u}(\mathcal{L}_2)$, respectively, which are topologically equivalent to two-dimensional cylinders. For each fixed value of $C \lesssim C_2$, these manifolds lie on the three-dimensional Jacobi integral surface $\tilde{J}^{-1}(C)$ (see Figure 3.8). When restricted to the invariant manifolds, the solutions spiral. The solutions of the restricted problem near L_2 for $C \lesssim C_2$ are of the form of the linearized flow as is verified from [64]

$$\mathbf{x}(t) = \alpha_1 \mathbf{v}_1 e^{\lambda t} + \alpha_2 \mathbf{v}_2 e^{-\lambda t} + 2Re(\beta e^{i\nu t} \mathbf{w}), \tag{3.35}$$

where $\mathbf{x} = (x_1, x_2, x_3, x_4), x_3 = \dot{x}_1, x_4 = \dot{x}_2$, and x_1, x_2 are the coordinates of (1.61) for the restricted problem in barycentric rotating coordinates. $\beta \in \mathbb{C}, \alpha_1, \alpha_2 \in \mathbf{R^1}; \mathbf{v}_1, \mathbf{v}_2 \in \mathbf{R^4}, \mathbf{w} \in \mathbb{C}^4$. $\mathbf{x}(t)$ in the form (3.35) is real.

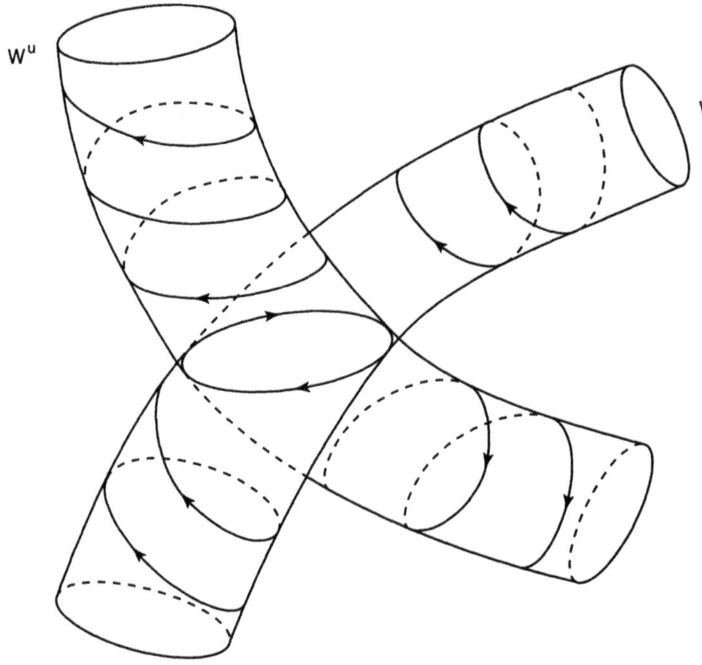

Figure 3.8 Local invariant manifolds to a Lyapunov orbit.

Using (3.35), some linearized solutions in a neighborhood of an orbit $\boldsymbol{\zeta}(t) \in \mathcal{L}_2(C), C \lesssim C_2$ are shown in Figure 3.9 in the *neck region* which is projected onto physical x_1, x_2 space. $\boldsymbol{\zeta}(t) = (\zeta_1(t), \zeta_2(t), \dot{\zeta}_1(t), \dot{\zeta}_2(t)) \in \mathbf{R^4}$. Vertical lines ℓ_1, ℓ_2 are chosen on either side of the Lyapunov orbit. Projected solutions on the projected invariant manifolds are shown.

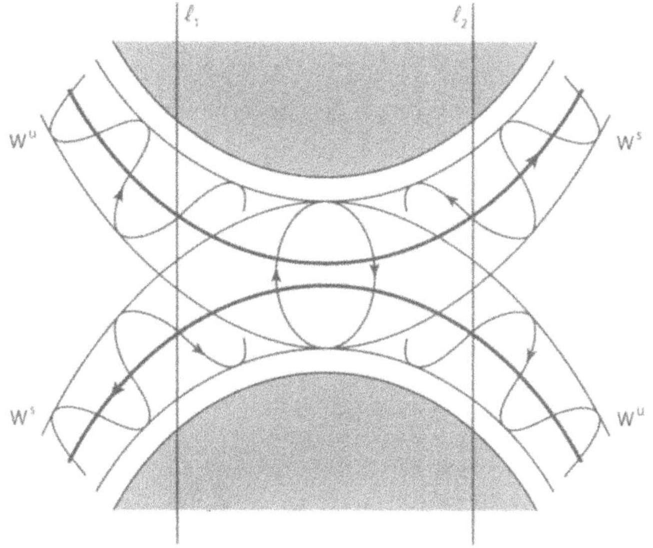

Figure 3.9 Transit orbits in neck region.

Definition 3.28 *Orbits with* $\alpha_1\alpha_2 < 0$ *are called* transit *orbits.*

Transit orbits cross the *equilibrium region* in the neck, lying between ℓ_1, ℓ_2, for t passing from $-\infty$ to ∞ or from ∞ to $-\infty$. They are shown in Figure 3.9, passing in either direction. They have the same appearance for the full restricted problem by continuity with respect to μ.

Lemma 3.29 (Conley) *Transit orbits exist in the circular restricted problem for* $C \lesssim C_2$ *in the neck region near* L_2.

The set of transit orbits in fact can be characterized in the full restricted problem. Let $\tilde{N}$ be the region bounded by the lines ℓ_1, ℓ_2 in the neck shown in Figure 3.9. By Theorem 3.27, $\tilde{N}$ is topologically equivalent to a region N which is the Hill's region in the neck for the full restricted problem. Let G be the subset of $\tilde{J}^{-1}(C)$ which projects onto N. It was proven in [64] by looking at the linearized Hamiltonian for the restricted problem that G is topologically equivalent to the product of a two-dimensional sphere and an open interval. Corresponding to ℓ_1 in the full restricted problem on $\tilde{J}^{-1}(C)$ is therefore a two-dimensional sphere, and likewise for ℓ_2. These are called bounding spheres. Thus, the set of transit orbits are characterized by those orbits which pass from one bounding sphere to another. Corresponding to ℓ_1, ℓ_2 we label these bounding spheres S_1^2, S_2^2, respectively.

We now assume $\mu \ll 1$.

Lemma 3.30 *A transit orbit in N is a ballistic capture transfer.*

Proof. Let $\varphi(t)$ be a transit orbit, where $C \lesssim C_2$. The part of the Hill's region to the right of the neck is $\mathcal{H}_E$ and the part of the region to the left of the neck is $\mathcal{H}_M$. Assume the transit orbit in G passes from the right to the left. Its initial point is $\varphi(0) \in S_2^2$, and its final point is $\varphi(t_F) \in S_1^2$. $\varphi(0)$ lies in the portion of the Hill's region about P_1 and $\varphi(t_F)$ lies in the portion of the Hill's region about P_2. Now, the function E_2 in rotating P_2-centered coordinates $\bar{x}_1, \bar{x}_2$ is given by (3.15). When $C = C_2$, in these coordinates, the distance of L_2 from P_2 is given by

$$r_{23} = \alpha + \mathcal{O}(\alpha^2), \tag{3.36}$$

$\alpha = (\mu/3)^{\frac{1}{3}}$. Thus, for $\mu \ll 1, r_{23} \approx \alpha$. At $L_2, \dot{\mathbf{x}} = 0$, and (3.15) yields

$$E_2 = \left(-3^{\frac{1}{3}} + \frac{1}{2}3^{-\frac{2}{3}}\right)\mu^{\frac{2}{3}} + \mathcal{O}(\mu^b),$$

where $b > \frac{2}{3}$. Thus, since $-3^{-\frac{1}{3}} + \frac{1}{2}3^{-\frac{2}{3}} = -1.20187\ldots$, then for $\mu \ll 1$,

$$E_2(L_2) < 0.$$

This implies for $C_2 - C \gtrsim 0$ sufficiently small and for $\varphi(t_F) \in S_1^2$ sufficiently near P_2 that

$$E_2(\varphi(t_F)) < 0.$$

This implies that the small arc $\varphi(t), t \in [0, t_F]$, is a ballistic capture transfer from $\varphi(0)$ to $\varphi(t_F)$ by Definition 3.13. □

It is remarked that in [64], Conley defined *capture orbits* for $C \lesssim C_2$ as those which asymptotically approach a Lyapunov orbit as $t \to +\infty$ or $t \to -\infty$. Such orbits would lie on $\mathbf{W^s}, \mathbf{W^u}$. This definition of capture corresponds to a type of permanent capture considered in section 3.1. However, it is different since if the orbit asymptotically approaches the Lyapunov orbit as $t \to \infty$, whereby $|\mathbf{Q}|$ is bounded, it need not have the property that $|\mathbf{Q}| \to \infty$ as $t \to -\infty$. In fact since $C \lesssim C_2$, the particle P_3 is trapped in the peanut-shaped Hill's region and therefore $|\mathbf{Q}| \not\to \infty$, as $t \to -\infty$. This implies that we have a definition of a different capture process.

Definition 3.31 *Let $\gamma(t)$ be in bounded periodic orbit in the three-dimensional elliptic problem in barycentric inertial coordinates $\mathbf{Q} = (Q_1, Q_2, Q_3)$, $\gamma(t+T) = \gamma(T), 0 < T < \infty$. Let $\varphi(t)$ be an orbit asymptotic to $\gamma(t)$ as $t \to \infty$. That is, as $t \to \infty, |\varphi(t) - \gamma(t)| \to 0$. Then we say $\gamma(t)$ is* asymptotically captured into the periodic orbit *$\gamma(t)$ as $t \to \infty$.*

This definition is similiarly valid for the planar restricted problem in rotating coordinates. As we saw, asymptotic capture into a periodic orbit in general need not imply permanent capture, nor ballistic capture since one can have $E_2(\gamma(t)) > 0$. In a neighborhood of the neck, asymptotic capture into a Lyapunov orbit implies ballistic capture by Lemma 3.30.

Consider Figure 3.10, with the coordinate system centered at P_1.

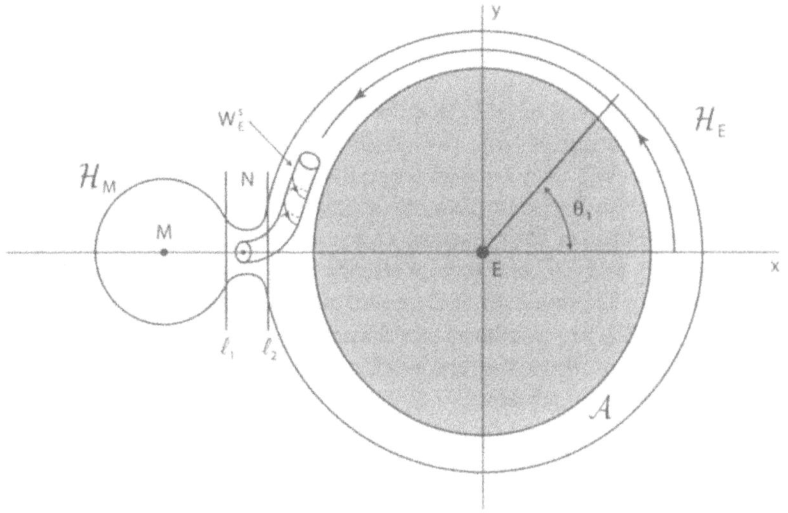

Figure 3.10 Proof of primary interchange capture by asymptotic capture to a Lyapunov orbit.

A Lyapunov orbit $\zeta(t)$ is shown corresponding to energy $C \lesssim C_2$, with an invariant manifold $\mathbf{W^s_E}$. The subscript E is used to identify that $\mathbf{W^s}(\zeta)$ lies in $\mathcal{H}_E$.

The local neck region about ζ is bounded by two lines ℓ_1, ℓ_2. Let x_{L_2} denote the location of L_2 on the x_1-axis, which is -1. The line ℓ_2 is chosen to be at the position $x_{L_2} + k\mu^{\frac{1}{3}}$ on the x_1-axis where $k \lesssim 3^{\frac{1}{6}}$. We take μ sufficiently small to insure the existence of $\mathbf{W^s_E}$ within N.

Theorem 2.3 implies the existence of a family of two-dimensional invariant tori in the region G_E on $\tilde{J}^{-1}(C)$ which projects into $\mathcal{H}_E$. They exist provided (2.7) is satisfied and they don't get too close to the point L_2, which is not in the set $\hat{\mathcal{D}}$, defined in Theorem 2.1, or too close to E. They will lie close to Z. McGehee proved in [155] that in the set G_E where $x > x_{L_2} + \mu^{\frac{1}{3}}k$ invariant tori exist. This is done for $C = C_2$. By continuity it is valid for $C \lesssim C_2$.

This leaves a thin annular region in $\mathcal{H}_E$ whose inner boundary is the outer boundary of the projected tori from G_E onto $\mathcal{H}_E$ and whose outer boundary is $Z \cup \ell_2$. We call this annular region A and it is shown in Figure 3.10. It has width $\mathcal{O}(\mu^{\frac{1}{3}})$. Let $\mathcal{A} \in \tilde{J}^{-1}(C)$ project onto A.

Lemma 3.32 *There exists infinitely many primary interchange capture transfers* $\mathbf{b}(t) \in \mathcal{A}$ *satisfying the following: (i)* $\theta_1 = 0$ *for* $t = 0$, *(ii)* $\mathbf{b}(t)$ *are asymptotic to* $\boldsymbol{\zeta}(t)$ *as* $t \to \infty$, *(iii)* $\mathbf{b}(t)$ *lies near* Z *in position space, for each* $C \lesssim C_2, \mu \ll 1$.

Proof. McGehee proved in [155] that for any trajectory moving in $\mathcal{A}$, $\dot{\theta}_1(t) > 0$. This means in negative time $\tau = -t, d\theta_1/d\tau < 0$. This implies that in the variable τ, when $\mathbf{W^s_E}$ is extended beyond ℓ_2, it will move in the projected region A counterclockwise in backwards time until $\theta_1 = 0$. Thus, switching back to t, the orbits on $\mathbf{W^s_E}$ starting at $\theta_1 = 0$, where we assume $t = 0$, will spiral along $\mathbf{W^s_E}$ and become asymptotically captured by $\boldsymbol{\zeta}(t)$ as $t \to \infty$. By the same proof in Lemma 3.30, it is proven also that $E_2(\boldsymbol{\zeta}(t)) < 0$. Thus, the trajectories on $\mathbf{W^s_E}$ are primary interchange capture transfers by Definition 3.20 since as $t \to \infty, \theta_1 = \pi$ is achieved, and actually infinitely many times due to the asymptotic spiraling. □

Remark. It is seen that there exist orbits in $\mathcal{A}$ lying near $\mathbf{W^s_E}$ which when they cross the line ℓ_2, don't asymptotically approach $\boldsymbol{\zeta}(t)$ and are transit orbits in N. Thus, primary interchange capture transfers which arrive near $\boldsymbol{\zeta}(t)$ in N in finite time exist as well.

It turns out that the dynamics of trajectories in $\mathcal{A}$ is quite complicated and gives rise to orbits which cycle about E as many times as desired, pass through N into $\mathcal{H}_M$, and likewise perform any given number of cycles about M, where they again pass through N into $\mathcal{H}_E$, etc. Also, orbits with a predetermined cycling pattern about E, M ending (and/or beginning) in asymptotic capture exist.

A numerically assisted proof of this is given in [136]. The idea of the proof is to first show that there exist transversal homoclinic orbits lying on $\mathbf{W^s_E}$ for a set of values C, μ, with $C < C_2$ and $\mu \ll 1$. Recall that in section 2.3, we defined homoclinic points for two-dimensional area-preserving maps. When a transverse homoclinic point occurs due to the breaking of a homoclinic loop, the dynamics is complicated. Analogous to this, for flows, we define in general *homoclinic orbits* associated to an equilibrium point $\mathbf{O}$ or a periodic orbit $\boldsymbol{\phi}$.

Definition 3.33 *A* homoclinic orbit *associated to* $\mathbf{O}$ *or* $\boldsymbol{\phi}$ *is an orbit tending to* $\mathbf{O}$ *or* $\boldsymbol{\phi}$ *as* $t \to \pm\infty$. *It therefore lies on the invariant manifolds associated* *to* **O** *or* ϕ. *A homoclinic orbit is called* transversal *if at some point* **q** *of the orbit the tangent spaces to the stable and unstable manifolds at* **q** *span the full tangent space. If the manifolds are one-dimensional the existence of a homoclinic orbit implies that both manifolds coincide with the homoclinic orbit by uniqueness.*

When $\mathbf{W_E^s}$ coincides with $\mathbf{W_E^u}$, then the orbits on $\mathbf{W_E^s}$ are homoclinic orbits. Values of μ, C are shown to exist where this occurs [136]. Geometrically, on G_E for a fixed value of μ, C the invariant manifold $\mathbf{W_E^s}$ lies in $\mathcal{A}$ and the orbits on the manifold asymptotically approach the Lyapunov orbit $\boldsymbol{\zeta}(t)$ as $t \to \pm\infty$ in the neck. It just appears as a cylindrical tube T looping around E. If a two-dimensional surface of a section $\sum$ is made for T on $\tilde{J}^{-1}(C)$, one obtains a topological circle. We call the invariant manifold $\mathbf{W_E^s}$ in this case a *homoclinic tube.* The tangent space to points on orbits on the tube don't span the full three-dimensional tangent space of $\tilde{J}^{-1}(C)$ at these points. They only span the two-dimensional tangent space to the invariant tube at these points.

If at a point on the tube, the tangent space to the homoclinic orbit spans the full tangent space, this means that the tube breaks and has a self-intersection. This breaking has the effect of producing a complicated dynamics for a two-dimensional map on a two-dimensional surface of section to T on $\tilde{J}(C)$. The dynamics for such a map is analogous to that shown in Figure 2.8. This map gives a way to prescribe infinitely many different possible motions that P_3 can have as it passes through the neck as mentioned above. It is called a *shift map.* The details of this are not presented in this section since we will present this type of approach using *symbolic dynamics* in section 3.6 for a different problem of relevance to the presentation of this book , where a map can be defined on a hyperbolic network also described by Figure 2.8.

It is remarked that a result related to that in [136] was obtained in [127] using similiar methods by examining the numerical existence of transverse heteroclinic orbits (defined similiarly as for homoclinic orbits) in the portion of the Hill's region about P_2 for special values of C, μ in $C < C_1$, $\mu \ll 1$.

3.3.2 Numerical Simulation of Practical Ballistic Capture

A numerical simulation of practical ballistic capture is carried out in [25]. The example represents a generalization of the type of application Conley proposed in [63], and it is done with substantially more realistic modeling. Conley discussed the possibility of constructing low energy transfers to the Moon from the Earth that would be useful for applications using the planar restricted problem. The value of the Jacobi constant is fixed in $C \lesssim C_2$ and the transfers would pass through the neck near L_2. They would represent minimal energy transfers since $C \lesssim C_2$. No actual examples were constructed [63]. As seen below, this is a nontrivial problem. It represents the first realistic numerical simulation of a ballistic capture transfer.

The model we use is the three-dimensional three-body problem, where the three bodies are P_1 = Earth(E), P_2 = Moon(M), and a particle P_3 of negligible mass. The orbits of the Earth and Moon are provided by the planetary ephemeris DE403. This is a database which accurately describes the realistic orbits of the planets of our solar system over time [212]. The orbits of the Earth and Moon are very close to circular but have a slight eccentricity, and the orbital elements do slightly fluctuate as a function of time due to nongravitational effects. So this is close to the three-dimensional circular restricted problem. We call this the three-dimensional pseudocircular restricted problem, PR3BP-3D. Because this problem is very close to the circular restricted problem, in our analysis below we will calculate the Hill's regions and the Jacobi constant for PR3BP-3D using the formula for the circular restricted problem to gain an approximate sense of the dynamics of motion of P_3.

The results of [25] are briefly described to show how the theoretical treatment of primary interchange transfers, general ballistic capture transfers to the weak stability boundary W, and transit orbits can be illustrated in a realistic setting. This presentation is more descriptive and is summarized here for brevity. For those interested in theoretical issues, this subsection can be skipped.

The transfer constructed in [25] is for a spacecraft P_3 with a very low ΔV capability using engines that if left on continuously would only be able to achieve 40 meters per second *per day.* It was desired to have P_3 transfer to M from $r_{13} = r_E + 200$ km. Because of the low ΔV capability, this problem motivated defining weak stability boundaries and ballistic capture transfers. Hohmann transfers were not possible since several hundred of meters *per second* would have to be applied instantaneously upon arrival at M.

In order for P_3 to arrive in ballistic capture relative to M at $\mathbf{Q_F}$ it was seen that arrival at a location between capture and escape as defined in the numerical algorithm for $\mathcal{W}$ would enable such a transfer to exist. However, this problem is not for the the planar restricted problem. It is for PR3BP-3D. The algorithm and definition for $\mathcal{W}$ need to be generalized to this case. We describe how this is done.

The generalization of $\mathcal{W}$ is labeled $\tilde{\mathcal{W}}$. $\tilde{\mathcal{W}}$ is five-dimensional. This is because the line ℓ used in the definition of $\mathcal{W}$ in Section 3.2 shown in Figure 3.5 in the plane is generalized to three dimensions and is now parameterized by two spherical angles, say $\theta_2 \in [0, 2\pi], \beta_2 \in [0, \pi]$ instead of only θ_2. β_2 is measured from the positive $\hat{Q}_3$-axis of a rotating coordinate system $\hat{Q}_1, \hat{Q}_2, \hat{Q}_3$ centered at the Moon and normal to the osculating E–M orbital plane. The $\hat{Q}_1, \hat{Q}_2$-plane is analogous to the rotating X_1, X_2-plane in Figure 3.5, and the rotation is now about the $\hat{Q}_3$-axis instead of the origin, where the $\hat{Q}_3$ remains invariant. Because DE403 is used, the $\hat{Q}_3$-axis, normal to the $\hat{Q}_1, \hat{Q}_2$-plane, slightly varies. θ_2 is the angle measured in the $\hat{Q}_1, \hat{Q}_2$-plane, relative to the $\hat{Q}_1$-axis. This coordinate system rotates with the motion of E, M and is not uniform. An analogous algorithm is defined as in section 3.2 and the initial velocity vector of P_3 is again propagated along the line ℓ. Let $t = \tilde{t}$ be the initial time of propogation. The initial velocity vector lies on a plane U perpendicular to ℓ. This is more general than the situation previously considered in section 3.2, where the velocity vector of P_3 on ℓ, with magnitude V_2, only pointed in either the direct or retrograde direction, for each value of θ_2. An angle $\alpha_2 \in [0, 2\pi]$ with respect to a coordinate system on U is used to parameterize the direction of the velocity vector. E_2 is computed by (3.12), where $e_2 \in [0, 1]$ is again a parameter and is held fixed along ℓ as before, where V_2 is adjusted with the same expression as before. Thus $e_2, \theta_2, \alpha_2, \beta_2, \tilde{t}$ are the parameters for $\tilde{\mathcal{W}}$ where $r^* = r^*(e_2, \theta_2, \alpha_2, \beta_2, \tilde{t})$ is computed defining $\tilde{W}$. $\tilde{t}$ is a parameter since the planetary ephemeris varies as a function of time. This implies that for different values of $\tilde{t}$ in the rotating coordinate system, although both E and M lie on the $\hat{Q}_1$-axis, their relative distance will change. Thus, $\tilde{W}$ is five-dimensional in a seven-dimensional extended phase space.

We now consider an inertial P_1-centered coordinate system. It was desired to have the spacecraft be captured at a point $\mathbf{Q}_F$ over the north lunar pole where $\beta_2 = 0$, $e_2 = 0$, and the velocity vector is pointed in the antiE–M line direction where $\alpha_2 = 0$ in the coordinate system used on U. Arbitrarily choosing $\theta_2 = 0$, it turns out that $r^* \cong 35{,}000$ km. By definition of $\tilde{\mathcal{W}}$, the solution $\varphi(t) = (Q_1(t), Q_2(t), Q_3(t), \dot{Q}_1(t), \dot{Q}_2(t), \dot{Q}_3(t))$ at the desired capture point $\mathbf{Q}_F$ on $\tilde{\mathcal{W}}$ at $t = t_F$ will give rise to a sensitive motion as $|t - t_F|$ increases from $t = t_F$. This means that negligibly changing $|\dot{\mathbf{Q}}(t_F)|$ at $\mathbf{Q}_F \equiv \mathbf{Q}(t_F)$ causes large deviations in $\mathbf{Q}(t)$ as $|t - t_F|$ increases.

This property is used to our advantage to find an initial point $\mathbf{Q}_0$ near E with a velocity vector $\dot{\mathbf{Q}}_0$ leading to $\varphi(t_F) \in \tilde{\mathcal{W}}$. This is accomplished by backwards integration. The solution $\varphi(t)$ with initial value $\varphi(t_F) = (\mathbf{Q}_F(t_F), \dot{\mathbf{Q}}(t_F)) \in \tilde{\mathcal{W}}$ is numerically integrated in backwards time $t < t_F$ until it achieves a periapsis relatively near E at $t = t_0$ defined by the condition $\frac{d}{dt}(|\mathbf{Q}(t)|) = 0$ at $t = t_0$. To obtain different periapsis points, the initial velocity vector for $t = t_F$ is slightly varied by a small constant vector $\boldsymbol{\delta}$; that is, we use as the initial velocity vector $\dot{\mathbf{Q}}_F(t_F) + \boldsymbol{\delta}$ and then perform another backwards integration. $\boldsymbol{\delta}$ is chosen so as to always have the same direction as $\dot{\mathbf{Q}}(t_F)$, and only $|\boldsymbol{\delta}|$ is varied. As $|\boldsymbol{\delta}|$ is varied, the different periapsis points at different final times $t_0(|\boldsymbol{\delta}|)$ are chosen so as to have minimum value of $r_{13} = |\mathbf{Q}(t_0)|$, while keeping $|\boldsymbol{\delta}|$ within a small tolerance.

Once a minimum value of r_{13} is found, this yields in forwards time a trajectory from $\mathbf{Q}(t_0) \equiv \mathbf{Q}_0$ to $\mathbf{Q}_F$ arriving at $\mathbf{Q}_F$ on the set $\tilde{\mathcal{W}}$, where $e_2 = 0$, and is therefore ballistically captured. This yields a weak ballistic capture transfer $\varphi(t)$ from $\mathbf{Q_0}$ to $\mathbf{Q_F}$. It was found that $r_{13} = 154,089$ km, and P_3 arrives at $\mathbf{Q_F}$ where $\Delta t = t_F - t_0 = 16$ days.

The dynamics of this transfer is interesting. The trajectory moves approximately 75,000 km out of the E–M orbital plane, then moves sharply down near to the neighborhood of L_2. The value of $C \lesssim C_2$. It then goes to a minimum approximately 100,000 km below the E–M plane, then moves sharply up over the north lunar pole to $\mathbf{Q}_F$. In Figure 3.11 a plot is shown in an inertial E-centered coordinate system projected on the E–S plane, S = Sun. Figure 3.12 is an out-of-plane view in a rotating coordinate system of Figure 3.11. This is a three-dimensional analog of a primary interchange capture transfer and is considerably more complicated. Moreover, P_3 moves beyond the neck region and considerably further to $\tilde{W}$ for ballistic capture.

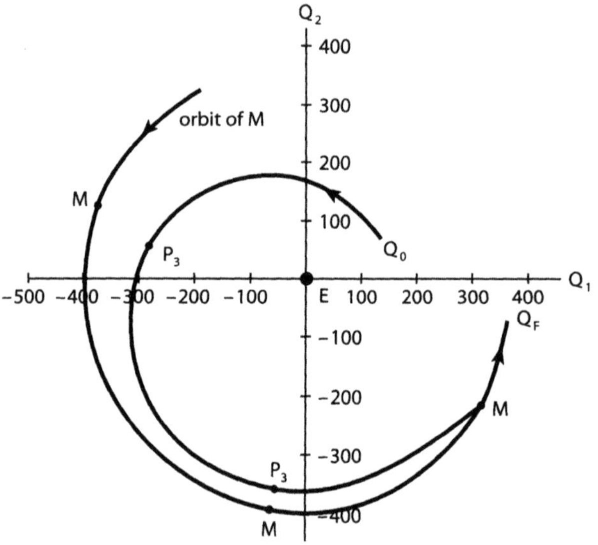

Figure 3.11 Projection onto Earth–Sun plane of ballistic capture transfer arc.

If this were a circular restricted problem, the trajectory $\varphi(t)$ is passing from the three-dimensional Hill's region about E to the Hill's region about M by passing through the narrow neck region. The Hill's regions shown in Figure 3.6 for the planar circular problem are generalized to the three-

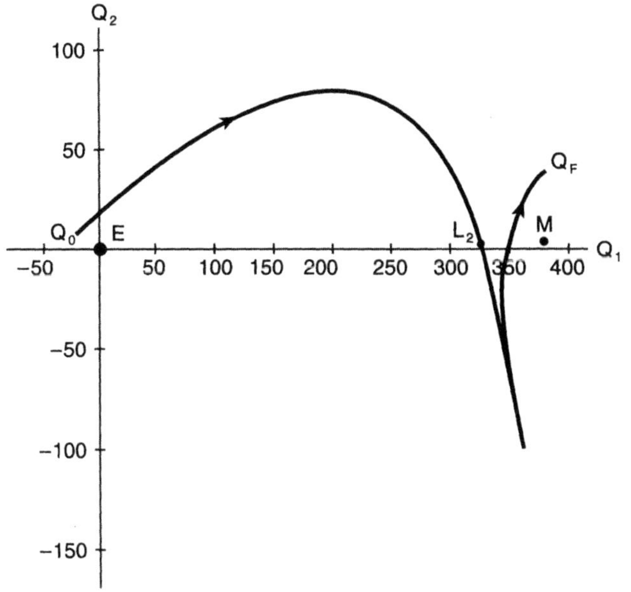

Figure 3.12 Projection onto plane normal to Earth–Sun plane in rotating coordinates.

dimensional circular problem in [219]. Thus, a three-dimensional neck is obtained.

Within the neck $\varphi(t)$ represents a three-dimensional version of a transit orbit. P_3 is passing from $\mathcal{H}_E$ to $\mathcal{H}_M$. When it enters $\mathcal{H}_M$ it moves to $\tilde{W}$ at $\mathbf{Q}_F$. Locally within the neck it is passing near to the stable manifold of the planar Lyapunov orbit as it enters the neck and near to the unstable manifold of the Lyapunov orbit as it exits the neck. This process is described in more detail in the next section for a related problem.

Now, the spacecraft has to reach $\mathbf{Q}_0$ with the correct velocity $\dot{\mathbf{Q}}(t_0)$ required for the ballistic capture transfer φ we just constructed. This is also a difficult problem and is solved by using the thrusting engines in a precise way. Due to the low thrust of the engines, once it is released at $r_{13} = r_E + 200$ km, it takes approximately *3,000* spirals about E to arrive at $(\mathbf{Q}_0, \dot{\mathbf{Q}}(t_0))$, where the time to do this is about *2 years*. During this time the engines can be properly thrusted so as to match the conditions at $\mathbf{Q}_0$. The engines are then shut off. It coasts to $\mathbf{Q}_F$ on the ballistic capture transfer and the engines are turned back on; it then slowly spirals down to a low circular orbit about the Moon where $r_{23} = r_M + 100$ km, taking about 120 days. This process is shown in Figure 3.13.

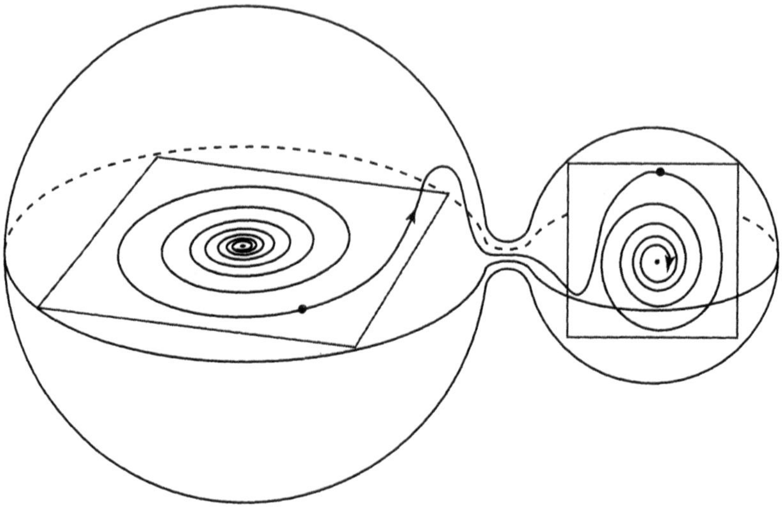

Figure 3.13 Representation of complete transfer from $\mathbf{Q_0}$ to low circular lunar orbit.

Since $|\mathbf{Q}_0| \gg r_E + 200$ km, and $|\mathbf{Q}_F| \gg r_M + 100$ km, this does not represent the desired solution to the capture problem. This is solved in the next section.

3.4 A LOW ENERGY LUNAR TRANSFER USING BALLISTIC CAPTURE

In this section a completely new type of ballistic capture transfer is generally described which represents the desired solution of the capture problem. It was discovered in order to resurrect a Japanese lunar mission and was operationally demonstrated in 1991. This is described in subsection 3.4.1. The design of this transfer uses weak stability boundaries in a way analogous to the example in subsection 3.3.2; however, in this case to achieve a general solution to the capture problem a four-body problem is used. The original design process is described briefly in subsection 3.4.1. This is an example of how a real application gave rise to a new type of capture process of theoretical interest.

We describe the dynamics of this trajectory heuristically using invariant manifolds and Hill's regions. This is accomplished by breaking up the four-body problem into two restricted three-body problems and is described in subsection 3.4.2.

The discovery of this transfer and the associated methodology of designing it represents a different approach to the design of transfer trajectories. This is elaborated on in subsection 3.4.3.

This is an applied section and as in subsection 3.3.2, where results are summarized, they are described in a general manner for brevity. This material presents a new type of ballistic capture transfer and illustrates some concepts previously presented. Historical comments are made.

3.4.1 Demonstration of A New Type of Lunar Transfer by A Spacecraft

In January 1990 Japan's ISAS Institute launched a pair of small spacecraft linked together into an elliptic Earth orbit. The smaller one, called *MUSES-B*, detached from the larger one. It was supposed to go to the Moon and into lunar orbit using a Hohmann transfer. The communication with *MUSES-B* was lost and its mission was not fulfilled. The larger craft, *MUSES-A*, was still in Earth orbit. It was to have been a communications relay for *MUSES-B*, and it also performed scientific experiments while in Earth orbit. It was then desired by the ISAS Institute to try to get *MUSES-A* to the Moon as a replacement for *MUSES-B*, and into lunar orbit with a desired lunar periapsis at capture of $r_{23} = r_M + 100$ km. *MUSES-A* had a very small ΔV capability of approximately 100 meters per second (m/s), far less than what is necessary to be placed into lunar orbit using a Hohmann transfer. This is because it was never designed to go to the Moon.

Guided by a methodology analogous to that described in subsection 3.3.2, a solution was found by Belbruno and Miller [38] at the Jet Propulsion Laboratory in June 1990 to enable *MUSES-A*, renamed *Hiten*, to reach the Moon on a ballistic capture transfer to the region $\tilde{\mathcal{W}}$. This transfer rescued the Japanese lunar mission. Without it, there was not enough ΔV to get *Hiten* to the Moon by any other means.

We now describe how this transfer was determined. The model is the same as in the previous example, except that the Sun's gravitational force was modeled, using DE403. This generalizes PR3BP-3D to a three-dimensional pseudocircular restricted four-body problem, PR4BP-3D. As in subsection 3.3.2 a backwards numerical integration from $\tilde{W}$ is performed at a given time where $r^* = r_{23} = r_M + 100$ km, implicitly implying that $e_2 = .95$, where $\theta_2, \alpha_2, \beta_2$ were prescribed. This produced a trajectory arc labeled II, shown in Figure 3.14. It went out to approximately four times the Earth–Moon distance, approximately 1.5 million kilometers from E, at its apoapsis $\mathbf{Q}_a$ with respect to E. Thus, arc II starting from $\mathbf{Q}_a$ represents a ballistic capture transfer to $\tilde{W}$. The time of flight on arc II is about 45 days. Arc

II could have been extended further in backwards time; however, it was stopped at the location $\mathbf{Q}_a$ to facilitate matching it with another arc.

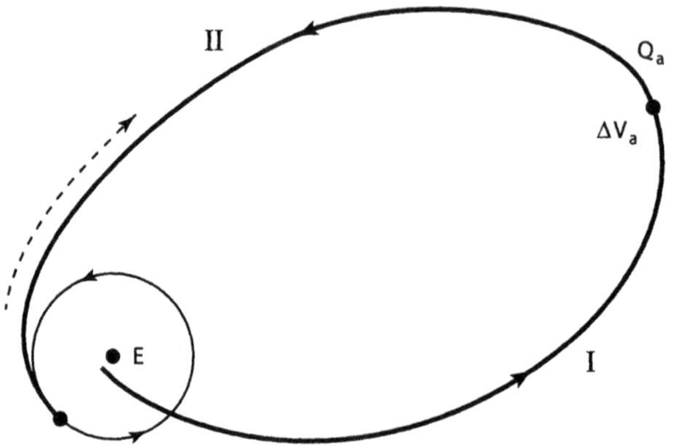

Figure 3.14 Procedure to obtain transfer using backwards integration from the stability boundary using connecting segments I, II.

Another arc, labeled I, was then generated from the *Hiten* spacecraft from its periapsis $\mathbf{Q}_0$ near E, where $r_{13} = 8{,}900$ km, to $\mathbf{Q}_a$ by performing a tiny maneuver of approximately 14 m/s at $\mathbf{Q}_0$ (see Figure 3.14). The time of flight on arc II is approximately 100 days. At $\mathbf{Q}_a$ a resulting mismatch of velocity ΔV_a occurred and after some effort was decreased to a small value of 30 m/s. A transfer was constructed whose total ΔV required approximately 44 m/s to get *Hiten* to ballistic lunar capture $\mathbf{Q}_F$, taking about 150 days. Thus $\Delta V_C = 0$. For a Hohmann transfer to arrive at the same value of e_2, approximately 200 m/s is required.

On April 24 of the next year *Hiten* left Earth orbit and successfully arrived at the Moon on October 2, 1991 on this new type of transfer, thereby enabling Japan to become the third nation in history to send a spacecraft to the Moon. The detailed construction of this transfer with the method just described is in [38, 39]. Expository and general nontechnical references include [75, 27, 84, 4].

The actual transfer *Hiten* used is shown in Figure 3.15 in an inertial E-centered coordinate system on the E–S plane, and is based on the transfer designed in [38] with negligible modifications. The small ellipse shown is an initial *phasing orbit* used to achieve the proper initial timing so that E, M, S are properly aligned.

The value of $\Delta V_a = 30$ m/s computed in [38] is very close to the actual value used in the flown transfer where $\Delta V_a = 34$ m/s. Hiten was originally targeted to ballistic capture at an altitude of 100 km but in flight it was slightly modified to arrive at a higher altitude of 72,422 km so as to slowly fly by M to perform additional scientific experiments in the E–M system while it flew near the equilateral Lagrange points L_4, L_5. This opportunity was possible due to the significant ΔV that was saved by the ballistic capture transfer, and the flexibility of this transfer due to its sensitive nature. *Hiten* later returned to the Moon to approximate ballistic capture in February 15, 1992 and was placed in lunar orbit where $\Delta V_C = 82$ m/s [221]. It was purposely crashed onto the lunar surface on April 10, 1993 [85].

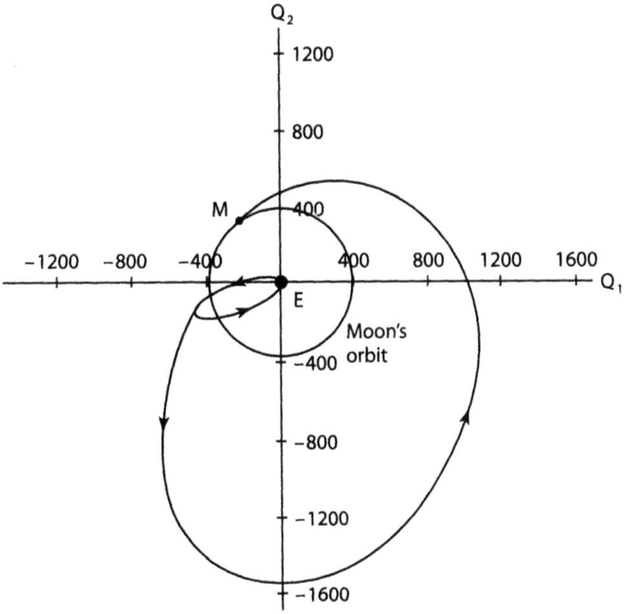

Figure 3.15 Ballistic capture transfer used by *Hiten* arriving at Moon on October 2, 1991.

The transfer just described, which goes from $\mathbf{Q}_0$ to ballistic capture at $\mathbf{Q}_F$, has $\Delta V_C = 0$. By comparison, a Hohmann transfer also leaving from the same point $\mathbf{Q}_0$ as *Hiten* would require approximately that $\Delta V_C = 200$ m/s (see section 3.1) and approximately the same value of ΔV_0 . The *Hiten* transfer requires 30 m/s for a midcourse maneuver ($\Delta V_a = 30$ m/s) for lunar targeting, whereas a Hohmann transfer does not. Therefore the total ΔV required for *Hiten after leaving* $\mathbf{Q}_0$ to achieve $e_2 = 0.95$ is 30 m/s, and the Hohmann case is 200 m/s. To make a meaningful comparison, we assume they both go to a circular lunar orbit, where $r_{23} = r_M + 100$. A ΔV of 648 m/s must be added to each. Thus, the total ΔV used by the Hohmann transfer to achieve circular lunar capture after leaving $\mathbf{Q}_0$ is 848 m/s, and the total ΔV required for the ballistic capture transfer is 678 m/s. This implies that the ballistic capture transfer yields a 20% improvement in ΔV

to achieve capture into a circular lunar orbit. The flight time of *Hiten* case is about 150 days and that of the Hohmann case is 5 days. This 20% figure is very significant for applications (See Remark 1.).

Remarks

Remark 1. Significant improvements to this type of ballistic capture transfer both in reducing the flight time to 90 days and in the elimination of ΔV_a have been made. In addition the value of r_{13} at $\mathbf{Q}_0$ can be adjusted to any value, and in particular to $r_{13} = r_E + 200$ km; see [34, 33, 42]. With these modifications the 20% improvement just calculated increases to 25%. For other ΔV comparisions of this transfer see [217].

Remark 2. Relating the ballistic capture transfer described here to the capture problem would require that the transfer be planar, and a pure elliptic motion is modeled for the motions of E, M, S. This type of transfer can also be obtained in this case, and an example is discussed in the next subsection. We can also extend the definition of the capture problem from the model ER4BP-2D to PR4BP-3D if desired in the natural way. For reference we refer to this transfer type as *exterior WSB-transfers*, where WSB represents weak stability boundary. The term exterior is used since P_3 approaches M from outside the distance of M to E. This compares to the one discussed in subsection 3.3.2, where we refer to *interior WSB-transfers* since they transfer to M interior to the distance from E to M. A survey of the exterior and interior transfers can be found in [42].

Remark 3. Both the exterior and interior transfers can be classified as bounded ballistic capture transfers, where we extend Definition 3.15 to include PR4BP-3D.

Remark 4. The backwards integration method is unwieldy to use, making it undesirable for applications, although that method was successfully used for *Hiten*. It is called the *backwards method.* A standard forward searching algorithm using differential correction can be used to find both exterior and interior transfers. This is called the *forward method.* It is described in [33, 34]. We elaborate on this. In applications it is required to match specific starting values at E at $\mathbf{Q}_0$. Let $\mathbf{V}_0$ be the initial velocity vector. The direction of $\mathbf{V}_0$ yields an initial inclination i_E with respect to the Earth's equator, as well as other parameters. i_E is an important parameter and is specified. Other prescribed constraints include t_0, r_{13}. It is found that when performing a backwards integration from $\tilde{W}$ at $\mathbf{Q}_F$, $|\mathbf{V}_F|$ or equivalently e_2 can be slightly varied to readily achieve the desired values of $r_{13} = |\mathbf{Q}_0|$. However, achieving t_0, i_E is difficult, especially due to the sensitivity of $\tilde{W}$. This is why the matching of arcs I, II is used in the backward method. In the forward method, no matching of arcs is generally required. A forward search algorithm which starts at the desired values of r_{13}, i_E, t_0 and other parameters with respect to E can be used to perform a targeted search using differential correction to the desired parameters at $\mathbf{Q}_F$. It turns out that only two need be required. They are r_{23}, i_M, where i_M is the inclination with respect to the Moon's equator. It turns out that specifying $e_2 < 1$ is not necessary, which is surprising. This is because the two control variables at $\mathbf{Q}_0$ turn out to be $\gamma_0, |\mathbf{V}_0|$. γ_0 is the flight path angle, which is the angle between $\mathbf{V}_0$ and $\mathbf{N}$ where $\mathbf{N}$ is the normal direction to $\mathbf{Q}_0$ in the plane spanned by $\mathbf{Q}_0, \mathbf{V}_0$. As long as $|\mathbf{V}_0|$ is chosen so that the trajectory has an apoapsis of approximately 1.5 million kilometers, and t_0 is suitably adjusted, then P_3 naturally goes to $\tilde{W}$ where $e_2 < 1$. It is generally found that γ is a small number. Using this algorithm, the full model PR4BP-3D can be used. It is a straight forward 2×2 method $|\mathbf{V}_0|, \gamma_0 \rightarrow r_{23}, i_M$. This is a vast simplification to the original backwards search method.

Other computational methods to compute the exterior transfer include using the global invariant manifolds associated with the Lyapunov orbits at L_1 for $C \lesssim C_1$ [126]. This method requires intermediate models, including the planar restricted problem where $C \lesssim C_1$. It turns out that for the exterior transfer, C can be substantially less than $3 < C_1$, so the restriction of $C \lesssim C_1$ would leave out a significant portion of the orbit space. A new method using genetic algorithms is described in [42], which may offer flexibility in searching the orbit space.

Remark 5. Let $\mathbf{\Phi}(t)$ represent the trajectory of an exterior transfer in three-dimensional physical space, where $\mathbf{\Phi}(0) = \mathbf{Q}_0, \quad \mathbf{\Phi}(t_F) = \mathbf{Q}_F$. Consider Figure 3.14. The way $\mathbf{\Phi}$ is able to achieve ballistic capture is physically motivated by the following description. First, by approaching $\tilde{W}$ from outside the lunar distance, it can start with any value of $|\mathbf{\Phi}(0)|$. This is a flexibility we didn't have in the interior case. When P_3 achieves its apoapsis at $\mathbf{Q}_a$ and it starts to fall back to the E–M system on arc II during its approach to M, the Sun's gravitational field slows P_3 down so that by the time it reaches the Moon's orbit it has a velocity that can match the Moon's. This action of S serves the role of eliminating the ΔV_C. This is seen in Figure 3.16, where prior to capture the point P_3 moves in approximate parallel formation with M for about a week. It is aligning itself with the motion of M. Relative to M, P_3 is nearly stationary prior to capture. To an observer on the Moon, a spacecraft would appear fixed in the sky.

This is also seen to be the case in the interior transfer. In that case, as P_3 approaches the neck where $C \lesssim C_2$, it moves very slowly, and it is moving in approximately parallel formation with M. With respect to M it is nearly fixed. It then passes through the neck for capture at $\mathbf{Q}_F$. This can be modeled by the restricted problem. In the exterior transfer, we can't readily approximate the motion of P_3 with the restricted problem. In the next section it is shown how to approximate this motion with the restricted problem, and analogous to the interior case, P_3 is nearly fixed outside the L_1 location near M prior to capture, and gets pulled through the neck of the Hill's region near L_1 to ballistic capture at $\mathbf{Q}_F$. Finally we note that at $\mathbf{Q}_a$ $\mathbf{\Phi}$ lies near a weak stability region $\tilde{\mathcal{W}}$ relative to E, S. The balancing of the accelerations there enable P_3 to approach $\mathbf{Q}_a$, then turn around for little or no ΔV to fall back to M. If the Sun's gravity were not modeled then approximately 200 m/s would be required to accomplish this at $\mathbf{Q}_a$.

3.4.2 Ballistic Capture Transfer, Invariant Manifolds, Hill's Regions

In this subsection we briefly summarize the analysis in [28]. A nominal exterior WSB-transfer is considered. It is shown in Figure 3.16. It is generated for the planar circular restricted four-body problem, CR4BP-2D, obtained from the model ER4BP-2D used to define the capture problem, by setting $e_{12} = e_{124} = 0$. This is sometimes referred to as a co-circular restricted four-body problem.

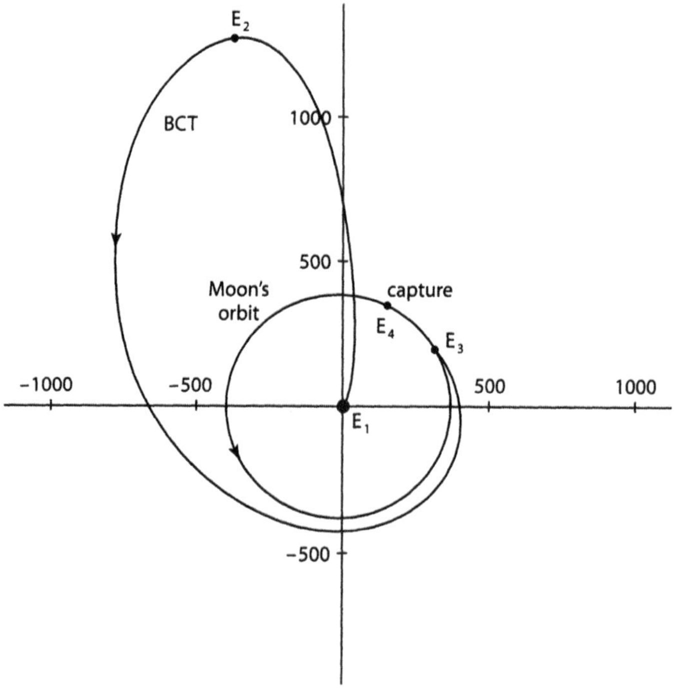

Figure 3.16 Nominal exterior WSB-transfer.

The particle P_3 starts at a distance $r_{13} = r_E + 185$ km. Its initial position $\mathbf{Q}_0$ is at a periapsis with respect to E. The initial velocity vector has magnitude 10.992 km/s, which yields an elliptic but nearly parabolic state with respect to E. P_3 arrives at an apoapsis point $\mathbf{Q}_a$ with respect to E at 3.65 times the mean distance $d(E, M)$ of E to M, assumed to be 384,000 km. The time to go from $\mathbf{Q}_0$ to $\mathbf{Q}_a$ is 35 days. At $\mathbf{Q}_a, \Delta V_a = 0$.

P_3 arrives 93 days after its start at $\mathbf{Q}_0$ to lunar periapsis at a point $\mathbf{Q}_F$ on the anti-E–M line, where $r_{23} = r_M + 100$ km. The magnitude of the velocity at $\mathbf{Q}_F$ with respect to M is 2.275 km/s. This implies $e_2 = 0.94$. $\mathbf{Q}_F \in \mathcal{W}$, where $\mathcal{W}$ is the weak stability boundary of the Moon if the gravitational field of S is ignored, we label $\mathcal{W}_M$.

For ease of analysis, we make yet another assumption: That CR4BP-2D can be broken up into two circular restricted three-body problems between E, M, P_3 and E, S, P_3, respectively. Breaking up the problem in this way yields sufficient accuracy for our estimates given the relatively short transit times on the transfer $\mathbf{\Phi}(t)$ we are now considering, where $\mathbf{\Phi}$ is viewed in position space in Figure 3.16. To break the four-body problem up we consider $\mathcal{W}_M$. $\mathcal{W}_M$ estimates a region where a particle is weakly captured by M. In physical space, beyond $\mathcal{W}_M$, the motion of P_3 is perturbed by M a negligible amount for time spans on the order of three months considered for $\mathbf{\Phi}(t)$. Thus, the maximal value of r_2^* defining $\mathcal{W}_M$ provides a measure of the maximal distance from M that the gravitational perturbation of M is felt in a significant way, for relatively short time spans. Beyond that distance from M, the modeling of the gravitational attraction of M can be ignored for short time spans.

This rationale provides a way to break up CR4BP-2D. To make this more precise, we first use W to approximate $\mathcal{W}_M$. This is more convenient since W can be easily computed and provides an analytic formula. It is verified by plotting W that

$$\max\{r_2^*(\theta_2, e_2, C)\}$$

occurs for $\theta_2 = \pi/2, 3\pi/2, e_2 = 1, C = C_1$ and has the value $\rho^* = 0.368 d(E, M)$ [30]. Define the disc D_o about E of radius

$$\rho = d(E, M) + \rho^*.$$

Let CR3BEM-2D, CR3BES-2D be the planar circular restricted problems between E, M, P_3 and E, S, P_3, respectively, for $\mu = 0.012, 0.000003$, respectively. Then, CR4BP-2D is broken up as follows:

$$CR4BP - 2D = \begin{cases} CR3BEM - 2D & \text{if } r_{13}(\mathbf{\Phi}) \leq \rho, \\ CR3BES - 2D & \text{if } r_{13}(\mathbf{\Phi}) > \rho. \end{cases}$$

Each of these two restricted problems has their own respective values of $\tilde{J}$. Let $C(E, M), C(E, S)$ be the associated values of $\tilde{J}$ for CR3BEM, CR3BEM, respectively.

To get a better understanding of the dynamics of $\mathbf{\Phi}$ we compute the relevant values of C along its path and examine the respective Hill's regions. For brevity we look at four points along $\mathbf{\Phi}$ labeled E_1, E_2, E_3, E_4, shown in Figure 3.16. More points could be chosen, however these give an adequate number to get an understanding of the variation of the Hill's regions.

E_1 is at $\mathbf{Q}_0$ at the start of $\mathbf{\Phi}$ at $t = 0$. E_2 is at $\mathbf{Q}_a$. E_3 is seven days prior to capture at $\mathbf{Q}_F$. E_4 is at Q_F at $t = t_F$.

We describe each case and refer to Figure 3.6 and Table 3.2.

Table 3.2 Values of Jacobi constants along $\mathbf{\Phi}(t)$.

	$C(E, M)$	$C(E, S)$
E_1	0.573	3.00230
E_2	2.287	3.00077
E_3	3.17466	3.00073
E_4	3.09210	2.99008

E_1 : $C(E, M) < 3$ so that P_3 can move throughout the plane in the local E, M-system. On the other hand $C(E, S) > C_2$ so P_3 is constrained to lie in a neighborhood of E.

E_2 : $C(E, S) = 3.000774 \lesssim C_1(E, S) = 3.000896$. The energy with respect to E in the E, S system has gone up (recall, $\tilde{J}$ is minus two times the energy of P_3). The energy at E_2 has gone up due to the Sun's attraction as Φ moved from $\mathbf{Q}_0$ to $\mathbf{Q}_a$. P_3 is more energetic now with respect to the E, S system. This is reflected in the fact that it's periapsis distance with respect to E has been raised.

The fact $C(E, S) \lesssim C_1$ means P_3 could escape E from the opening of the Hill's region near L_1. As P_3 falls towards M, the Sun is taking energy out of its motion with respect to the E, M system. The energy with respect to the E, M system has gone down, but P_3 can still move freely in the E, M system since $C(E, M) < 3$.

E_3 : $C(E, S)$ shows that the energy of P_3 in the E, S system has not changed appreciably and has gone down a negligible amount. The interesting change is in $C(E, M)$. This implies that the energy of P_3 in the E, M system has gone down significantly, and moreover,

$$C(E, M) = 3.17466 \lesssim C_1(E, M),$$

$C_1(E, M) = 3.184077$. Thus, the Hill's region in the E, M system has changed from where P_3 could move freely to where it is restricted to entering a neighborhood of P_2 except for a small opening near L_1. P_3 is seven days away from capture at $\mathbf{Q}_F$, and is moving outside the location of the open neck region near L_1.

In fact, it is nearly on the anti-E, M line at a distance $r_{23} = 69,818$ km. However, L_1 is at a distance of 64,105 km. It is verified that P_3 lies near the stable manifold to the Lyapunov orbit in the neck near L_1.

Over the next seven days it moves towards the neck, moving at first near the stable manifold $\mathbf{W^s}$ to the Lyapunov orbit $\boldsymbol{\zeta}$, following it into the neck near $\boldsymbol{\zeta}$, then moving near to the unstable manifold $\mathbf{W^u}$, to the other side of the neck and into the portion of the Hill's region about P_2 (see Figure 3.17).

It was conjectured in [28] that the invariant manifolds associated to L_1, L_2 for $C \lesssim C_1$ form a complex hyperbolic network in the Hill's region about P_2 consisting of the intersections of the hyperbolic manifolds associated to L_1, L_2 and that P_3 moves through this network to ballistic capture at $\mathbf{Q}_F$. Such a hyperbolic network was numerically shown to exist by [127] for selected values of C, μ. Although their value of μ was specific for the Sun, Jupiter system, it does imply that the hyperbolic network for μ for the Earth, Moon system is likely to exist.

E_4 : At $\mathbf{Q}_F$ the energy in the E, M system increases slightly and the neck opening gets larger, showing that P_2 can escape soon after capture. This indicates the instability of the capture. The energy with respect to the E, S system has not changed by very much, which is not surprising since the Sun's gravitational force is felt to a negligible amount for the time spans considered when P_3 is near the Moon.

It is noted that for $t > t_F$ the motion of P_3 is unstable, and over a few hours it will in general be ejected. A remarkable property of ballistic capture at W is that W in phase space is very thin in velocity magnitude. This means that if $P_3 \in W$ for $t = t_F$, then a tiny ΔV added to the velocity will cause it to be ejected in very short time. This works in reverse if a tiny ΔV is subtracted from the velocity. In this case it turns out that the capture discussed in the above example can be stabilized for only 20 m/s. The stabilized ellipse is slightly less eccentric and will stay for several months near this elliptic state [33, 34].

Koon et al. [126] followed a similar methodology as in [28], and additionally numerically investigated the global extensions of various manifolds and, as previously noted, numerically demonstrated the existence of a hyperbolic network about the smaller mass point P_2 in the restricted problem for specific values of C, μ. In the latter part of this chapter, without the aid of a computer we will *analytically* prove the existence of a different hyperbolic network about P_2 associated with ballistic capture for open sets of C, μ. The

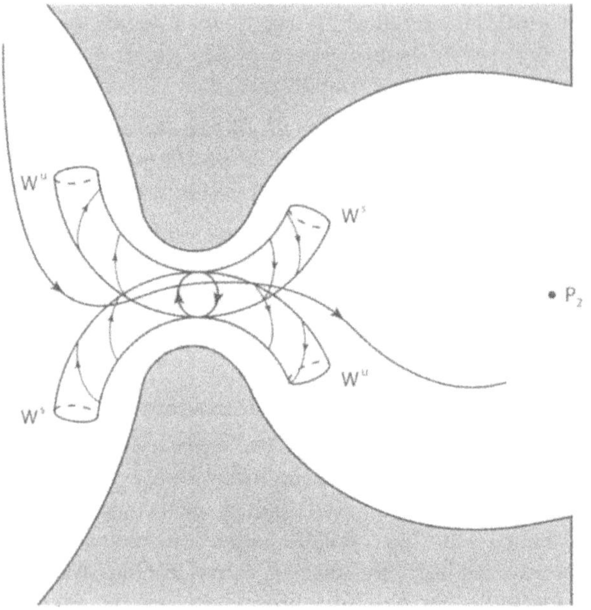

Figure 3.17 Capture process.

analytic nature of the proof provides insights into the capture process not previously available.

3.4.3 Historical Perspective and Concluding Remarks

The use of unstable regions such as the weak stability boundary and associated dynamics near the Moon provides a different methodology of computing transfers in the field of astrodynamics. The general field of mathematics used in this approach is given by dynamical systems theory. By comparison, the Hohmann transfer methodology designs high energy transfers (*high energy* and *low energy* are defined in section 3.1.5) which are Keplerian conic sections in nature. They are relatively stable. That is, if one is slightly perturbed, it will not change by very much. They are integrable in the sense that they can be explicitly solved for as solutions to the two-body problem. They approach the Moon with a value of $V_\infty \approx 1$ km/s. An example of transfers to the Moon which are a slight departure from the Hohmann transfer is provided by the *figure eight* lunar transfers used by the Apollo missions [219]. They are also as energetic as the Hohmann transfer, however they are nonintegrable since they exist as solutions of the three-body problem and cannot be explicity solved for by quadratures. Nevertheless, they are stable, since under small perturbation they change little.

The use of the weak stability boundary, invariant manifolds, and the unstable dynamics associated to it provided a numerical example described in subsection 3.3.2 of a new type of lunar transfer where the capture ΔV is zero. The exterior WSB-transfer described in Section 3.4 is the first low energy transfer designed by methodically exploiting the underlying unstable dynamics to be demonstrated by a real spacecraft. Unlike the Hohmann transfer, it is low energy, unstable, and nonintegrable. In this sense, *Hiten* represents a significant departure from the usual way transfers are designed. The fact that it approaches the Moon slowly and is nearly stationary with respect to it for about two weeks prior to capture is a completely different type of motion from the Hohmann transfer which arrives relatively quickly. There is a very short time interval for firing the rocket engines in order to be captured for a Hohmann transfer. Thus, if the firing is not done correctly, the spacecraft could be lost. The slow arrival of a ballistic capture transfer is actually much better for operations and for safety. The issues of having a critical short time interval to slow down in order to be captured into lunar orbit are not necessary with the exterior transfer. It can be placed gently into lunar orbit with negligible energy. The sensitive dynamics of this transfer make it very easy to do small correction maneuvers. The 25% savings in ΔV required for capture into a circular lunar orbit implies that less propellant is required. In fact, it can be shown that under various assumptions this propellant reduction can result in the total reduction of the mass of the spacecraft by one-half. Thus, a much less massive spacecraft could be designed at a lower cost and possibly a smaller launch vehicle could be used, which would further lower the cost. The exterior and interior transfers are also designed in a methodical manner utilizing the dynamical properties of the weak stability boundary and the invariant manifold structure nearby associated to the Lyapunov orbits. As will be seen in the last section, the weak stability boundary has a very complicated invariant manifold structure as well.

It is noted that in order to obtain a low energy transfer, such as the exterior lunar transfer, the time of flight relative to the Hohmann transfer is considerably longer. However, for robotic spacecraft the longer flight time of 90 days should not be an issue. It cost approximately one million dollars to place one pound into lunar orbit or on the lunar surface. Since an exterior ballistic capture transfer can place approximately twice as much mass on the lunar surface as compared to a Hohmann transfer for the same cost, then the exterior-type transfer would be very cost effective, if, for example, a lunar base were to be constructed.

The use of ballistic capture is not restricted to the Moon. A ballistic capture transfer to the weak stability boundary of Jupiter's moon Europa was designed in [218]. It is noted that the planned *SMART1* lunar mission of the European Space Agency was based originally on the interior transfer in [25], described in subsection 3.3.2. An exterior-type transfer is planned for Japan's *Lunar A* mission [124].

Another type of transfer from the Earth–Moon system utilizing ballistic ejection described in [26] has promising applications, and a Mars mission called *Planet B* of Japan is utilizing aspects of this process [123].

It is interesting to note that the original idea for the salvage of the Earth orbiting Hughes *HGS-1* satellite in 1998 was to use an exterior ballistic capture transfer to bring it to the Moon to effect a desired Earth inclination change. A different type of transfer was ultimaletly used instead as a modification of this idea [40]. Also, it is noted that of the four lunar missions in the 1990s, the exterior ballistic capture transfer was used by *Hiten*, and played a key role with *HGS-1*. The other two lunar missions using Hohmann transfers were *Clementine* and *Lunar Prospector* of the United States.

It is remarked that the *ISEE-3* mission in the late 1970s represented a departure from using stable motions when it orbited the Earth–Sun Lagrange point L_2 in a halo orbit, which is a periodic orbit about L_2 in three dimensions; the family of halo orbits bifurcate from the planar Lyapunov orbits [82, 81]. In general, halo orbits are unstable. The methodology of this design did not strive to use the unstable nature of the dynamics to minimize the ΔV to very low values in order to keep the orbit of the spacecraft near the nominal halo orbit. Numerically, this was done in [206], where the invariant manifold structure near the halo orbits was exploited. This has been further studied in [117] and is being applied to the *Genesis* mission of NASA.

There are likely many more interesting dynamics yet to be discovered associated with ballistic capture and escape near the weak stability boundary. The main result in the last section implies that this is the case. We will rigorously prove in section 3.6 that the dynamics associated to ballistic capture on W are not only unstable but chaotic.

3.5 PARABOLIC MOTION, HYPERBOLIC EXTENSION OF W

Parabolic motion in the planar circular restricted problem and the planar three-body problem is discussed. Motion on and near W, which is parabolic or *slightly hyperbolic* with respect to P_2, is shown to be parabolic with respect to P_1 under certain conditions. Results by Easton [76], Robinson [195] on the existence of parabolic orbits in the planar three-body problem are summarized.

Recall that parabolic and hyperbolic motion is defined in Definitions 3.6, 3.7, respectively. We now no longer add the constant term c_0 to $\tilde{J}$ that was assumed in subsection 3.2.2. $\tilde{J}$ is therefore given by (1.63) in barycentric rotating coordinates $\bar{x} = (x_1, x_2)$, for the planar circular restricted problem (1.61).

The Jacobi integral in barycentric inertial coordinates $\mathbf{Q} = (Q_1, Q_2)$ is given by (1.58), associated to the system of differential equations (1.55). For notational purposes we let $\hat{J}$ denote the Jacobi integral (1.58) in inertial coordinates, $\hat{J} = \hat{J}(\mathbf{Q}, \dot{\mathbf{Q}}, \mu)$. The Jacobi integral in barycentric rotating coordinates is given by (1.63), $\tilde{J} = \tilde{J}(\mathbf{x}, \dot{\mathbf{x}}, \mu)$.

3.5.1 Jacobi Integral Values for Parabolic Orbits for $\mu = 0$

We are interested in special orbits for the planar circular restricted problem which are parabolic with respect to P_1 for $\mu = 0$, and the corresponding value of the Jacobi integral. It is convenient to work in inertial coordinates. Also, it is convenient to set

$$J = -\frac{1}{2}\hat{J} = \frac{1}{2}|\dot{\mathbf{Q}}|^2 - \Omega - c, \tag{3.37}$$
$$\Omega = \frac{1-\mu}{r_{13}} + \frac{\mu}{r_2}, \quad c = Q_1\dot{Q}_2 - Q_2\dot{Q}_1.$$

Set $\mu = 0$. Thus, P_1 is at the origin and P_2 has mass $m_2 = 0$. We consider parabolic orbits for P_3 with initial conditions on the Q_1-axis for $t = 0$ with respect to P_1 given by

$$\mathbf{Q}(0) = (1, 0), \quad \dot{\mathbf{Q}}(0) = (0, \pm\sqrt{2}), \tag{3.38}$$

which are at periapsis.

The velocity vector is normal to the Q_1-axis. The + sign means it points up and the orbit moves in a direct fashion. If it is negative, it moves down in a retrograde fashion. We call these *direct, retrograde* parabolic orbits for $\mu = 0$.

It is *important to note* that the constant values of $J = \tilde{C} = -\frac{1}{2}C$. For notation we will label $\tilde{C}$ as C, not to be confused with the former value $\hat{J}$. Thus, in this new scaling of the Jacobi constant, the new value of C_1 is $-\frac{1}{2}$ times the previous value. Therefore on the set W, we now require the condition $C > C_1$. This scaling is assumed for the remainder of this book.

Lemma 3.34 *For parabolic orbits for $\mu = 0$ with respect to P_1,*

$$J = C = \mp\sqrt{2}, \tag{3.39}$$

where the upper sign(-) is for direct parabolic orbits, and the lower sign(+) is for retrograde parabolic orbits.

Proof. Equation (3.38) implies for $\mu = 0$ that

$$J = \frac{1}{2}(\pm\sqrt{2})^2 - \frac{1}{1} - (1 \cdot (\pm\sqrt{2})).$$

Thus $J = \mp\sqrt{2}$. □

The initial condition (3.38) of these parabolic orbits is at the distance of P_2 from P_1.

We assume now for convenience that the location of P_2 is at the location of $\mathbf{Q}(0) = (1-\mu, 0)$ when $t = 0$. This represents a reflection $x_1 \to -x_1$ in the restricted problem (1.61). We use the same variables x_k, y_k which now represent reflected coordinates.

It is remarked that if we considered initial parabolic conditions for P_3 for $\mu = 0$ on the Q_1-axis for arbitrary $Q_1(0)$, with normal velocity vector $(0, \dot{Q}_2(0))$, then $|\dot{Q}_2(0)| = \sqrt{2|Q_1(0)|^{-1}}$. In this case, we obtain more generally that $J = -Q_1(0)(\pm\sqrt{2|Q_1(0)|^{-1}})$. This implies that as $|Q_1(0)| \to \infty$, then $|J| \to \infty$.

3.5.2 Jacobi Integral Values for Parabolic Orbits with Respect to P_2

We assume that $\mu \neq 0$ until further notice. It is convenient to work in P_2-centered rotating coordinates $\bar{x}_1, \bar{x}_2$ defined by $\bar{x}_1 = x_1 - (1-\mu), \bar{x}_2 = x_2$ (see 3.16). $\tilde{J}$ is transformed into

$$\bar{G} = -|\dot{\bar{\mathbf{x}}}|^2 + (\bar{x}_1 + 1 - \mu)^2 + \bar{x}_2^2 + 2\left(\frac{1-\mu}{r_{13}} + \frac{\mu}{r_{23}}\right), \tag{3.40}$$

where $r_{13}^2 = (\bar{x}_1 + 1)^2 + \bar{x}_2^2, r_{23} = |\bar{x}|$. P_1 is now at the position (-1,0). Let P_3 have a *two-body* parabolic initial condition at its periapsis with respect to P_2, at a distance $r_{23} \equiv r_2$. This initial condition $(\bar{\mathbf{x}}(0), \dot{\bar{\mathbf{x}}}(0)) \in W$ for $e_2 = 1$,

$$\bar{\mathbf{x}}(0) = (r_2, 0), \quad \dot{\bar{\mathbf{x}}}(0) = \left(0, \pm\sqrt{2\mu r_2^{-1}} - r_2\right), \tag{3.41}$$

$+, -$ are for direct, retrograde orbits with respect to P_2, respectively. Thus,

$$\bar{G} = C = -\left(\pm\sqrt{\frac{2\mu}{r_2}} - r_2\right)^2 + (r_2 + 1 - \mu)^2 + 2\left(\frac{1-\mu}{r_{13}} + \frac{\mu}{r_2}\right). \tag{3.42}$$

This implies

$$\bar{G} = C = \mp 2r_2\sqrt{\frac{2\mu}{r_2}} - r_2^2 + (r_2 + 1 - \mu)^2 + \frac{2(1-\mu)}{r_{13}}.$$

Using the same normalization as for $\hat{J}$ in (3.37), we set

$$\tilde{G} = -\frac{1}{2}\bar{G},$$

and it is verified that, using $r_{13} = 1 + r_2$,

$$\tilde{G} = C = \bar{+}\, r_2\sqrt{\frac{2\mu}{r_2}} - \frac{1-\mu}{1+r_2} - r_2(1-\mu) - \frac{1}{2}(1-\mu)^2. \qquad (3.43)$$

Equation (3.43) yields the value of C on W for $e_2 = 1, \theta_2 = \pi$ in (3.29), where (3.29) is multiplied by 1/2.

For $\mu = 0, r_2 = 0$, the initial condition (3.41) in physical space is at the same location as (3.38) in physical space, and

$$\tilde{G} = C = -1.5. \qquad (3.44)$$

Thus, we have the following lemma.

Lemma 3.35 *At the location of P_2 for $\mu = 0$, the Jacobi energy C given by (3.44) is less than the values of C given by (3.39).*

Now, by continuity, for $0 < \mu \ll 1, 0 < r_2 \ll 1$, (3.44) implies for $r_{13} = 1 + r_2$,

$$\tilde{G} = C = -1.5 + \mathcal{O}(\mu) + \mathcal{O}(r_2) \approx -1.5. \qquad (3.45)$$

Also, for fixed $0 < r_2 \ll 1$, by continuity with respect to $0 < \mu \ll 1$ sufficiently small, (3.39) implies with $\mathbf{Q}(0) = (1+r_2, 0), \dot{\mathbf{Q}}(0) = (0, \sqrt{2/(1+r_2)})$,

$$J = C = \pm\sqrt{2} + \mathcal{O}(\mu) + \mathcal{O}(r_2) \approx \pm\sqrt{2}. \qquad (3.46)$$

This proves, the following lemma.

Lemma 3.36 *The value of C for two-body parabolic ($e_2 = 1$) initial conditions (3.41) with respect to P_2 on W for $0 < r_2 \ll 1, 0 < \mu \ll 1$, is $C \approx -1.5$. This value of C for each fixed $r_2, 0 < r_2 \ll 1$, and for $0 < \mu \ll 1$ sufficiently small is less than the value of $C \approx \pm\sqrt{2}$ for orbits with two-body parabolic initial conditions (3.38) with respect to P_1 in the restricted problem at a distance r_2 from P_2 on the x_1-axis, $r_{13} = 1 + r_2$.*

The weak stability boundary state for $e_2 = 1$ under our assumptions is therefore *less energetic* than for parabolic orbits with respect to P_1. Matching $|C|$ to $\sqrt{2}$ is necessary for section 3.6.

3.5.3 Matching Jacobi Energies

It is possible to choose μ so that on W for $e_2 = 1$ under the above conditions, $C = -\sqrt{2}$.

Lemma 3.37 *For initial conditions (3.41) on* W, *for* $e_2 = 1$,

$$\tilde{G}|_{r_2=0} = -\sqrt{2} \tag{3.47}$$

for

$$\mu = \frac{5}{4}\left(1 - \sqrt{1 - \left(\frac{4}{5}\right)^2 \left(\frac{3}{2} - \sqrt{2}\right)}\right) \approx 0.035. \tag{3.48}$$

Thus, by continuity, $\tilde{G} \approx -\sqrt{2}$ *for* $0 < r_2 \ll 1$.

Proof. Set $r_2 = 0$ in (3.43), yielding

$$\tilde{G}|_{r_2=0} = -1 + \mu - \frac{1}{2}(1-\mu)^2 = -(1-\mu)\left(\frac{3}{2} - \mu\right). \tag{3.49}$$

Setting $-(1-\mu)(\frac{3}{2} - \mu) = -\sqrt{2}$ yields (3.48). □

It is desired to match C for arbitrary $\mu \ll 1$. Thus the value of μ given by (3.48) may not be of interest. In order to match the values of C for arbitrary small μ and for r_2, it is necessary to extend the definition of W. In the retrograde case for $C = \sqrt{2}$ it is verified that no positive value of μ can be found to match the Jacobi energies. Since we need to match $|C| = \sqrt{2}$ in general and for arbitrary $\mu \ll 1$, another approach is used where the defintion of W is extended in subsection 3.5.5.

3.5.4 General Parabolic Orbits

In the previous subsections in this section, solutions were prescribed with two-body parabolic initial conditions relative to P_1 or P_2. This does not imply that the resulting trajectory is parabolic in the sense of Definition 3.6. That is, P_3 need not even become unbounded as $t \to \infty$.

The existence of true parabolic solutions in the planar three-body problem, of which the circular restricted problem is a special case, is considered.

Easton proved the existence of parabolic orbits in the planar three-body problem [76]. He accomplished this by constructing an invariant manifold at infinity, which is an invariant three-dimensional sphere S^3. To understand the meaning of parabolic orbits in this case we need to generalize Definition 3.6, which defined a parabolic orbit for the particle P_3 in the elliptic problem. In the planar three-body problem the situation is more general; however, it is analogous to Definition 3.6. We recall that in section 1.2 of chapter 1 where the planar three-body problem is defined for three particles P_1, P_2, P_3 using Jacobi coordinates $\mathbf{Q} \in \mathbf{R}^2, \mathbf{q} \in \mathbf{R}^2$. $\mathbf{q}$ is the relative position vector between P_1, P_2, and $\mathbf{Q}$ is the position vector of P_3 relative to the center of mass of P_1, P_2.

If $|\mathbf{q}(t)|$ is bounded for $t \geq 0$ and $|\mathbf{Q}(t)| \to \infty, |\dot{\mathbf{Q}}(t)| \to 0$, then the orbit $(\mathbf{q}(t), \mathbf{Q}(t))$ is called *ω-parabolic.* If these conditions are satisfied for t replaced by $-t$, then the orbit is called *α-parabolic.* If the energy of the full system between P_1, P_2, P_3 is $h < 0$, then for an ω-parabolic (α-parabolic) orbit, the binary system left behind between P_1, P_2 has negative two-body energy h as $t \to \infty (t \to -\infty)$.

The invariant sphere S^3 has associated to it stable and unstable manifolds. The parabolic orbits asymptotically approach S^3 as $t \to \pm\infty$ on these manifolds. The invariant sphere is not hyperbolic, so the stable manifold theorem we discussed in section 2.3 cannot be applied to this situation. Instead a technique is generalized from one developed by McGehee [150]. This method is applied by Easton and McGehee [79] for a different model problem. Easton applied this to the planar three-body problem for the case of the invariant sphere S^3, which is foliated by periodic orbits. He proved that the stable and unstable manifolds associated to this sphere have the smoothness of a Lipschitz manifold. Robinson proved that these manifolds to S^3 are real analytic [195].

These results are applied to the restricted three-body problem in the next section. We summarize them as

Theorem 3.38 (Easton, Robinson) *The ω- and α-parabolic orbits in the planar three-body problem are asymptotic to an invariant three-dimensional sphere S^3 at infinity, $|\mathbf{Q}| = \infty$, as $t \to \pm\infty$. These parabolic orbits lie on real analytic invariant manifolds associated to S^3. S^3 is foliated by periodic orbits.*

In the case of the planar circular restricted problem, the three-dimensional invariant sphere becomes a single periodic orbit, which is discussed in the next section. The case of a single periodic orbit at infinity is analogous to what happens in Sitnikov's problem. The single periodic orbit can be thought of as a one-dimensional invariant sphere.

3.5.5 Hyperbolic Extension of W

We extend the definition of W for the case $e_2 > 1$ for hyperbolic conditions with respect to P_2. By doing this we can achieve $|C| = \sqrt{2}$ for μ, r_2 arbitrarily small. Set $\tilde{B} = \{r_2 > 0, \theta_2 \in [0, 2\pi], \dot{r}_2 \in \mathbf{R}^1, \dot{\theta}_2 \in \mathbf{R}^1\}$.

Definition 3.39 *The* hyperbolic weak stability boundary W_H *corresponds to extending (3.22) to include the set*

$$\tilde{\Sigma}^* = \left\{(r_2, \theta_2, \dot{r}_2, \dot{\theta}_2) \in \tilde{B} | E_2 = \frac{\mu}{2}\left(\frac{e_2 - 1}{r_2}\right) > 0\right\}$$

for $e_2 > 1$, *and where (3.25) is the same with* $e_2 > 1$, *and (3.26) is the same, implying*

$$W_H = \mathcal{J}^{-1}(C) \cap \tilde{\Sigma}^* \cap \sigma^*. \tag{3.50}$$

C *is given by (3.29) with* $e_2 > 1$, *and where* $C > C_1$ *in the new scaling.*

We now calculate the value of $\tilde{G} = C$ restricted to W_H for special initial conditions.

Lemma 3.40 *Let*

$$\bar{\mathbf{x}}(0) = (r_2, 0), \quad \dot{\bar{\mathbf{x}}}(0) = \left(0, \pm\sqrt{\frac{\mu(e_2+1)}{r_2}} - r_2\right) \tag{3.51}$$

be hyperbolic initial conditions with respect to $P_2, e_2 > 1$. *Then, following the construction of (3.43), we substitute (3.51) into* $\tilde{G}$, *which implies that* $\tilde{G} \equiv \tilde{G}_H$ *is given by*

$$\begin{aligned} \tilde{G}_H = C = & \frac{1}{2}\frac{\mu(e_2 - 1)}{r_2} \\ & \mp r_2\sqrt{\frac{\mu(e_2+1)}{r_2}} - \frac{1-\mu}{1+r_2} - r_2(1-\mu) - \frac{1}{2}(1-\mu)^2. \end{aligned} \tag{3.52}$$

Proof. Proof is by direct substitution. □

Equation (3.52) is equivalent to (3.29) for $\theta_2 = \pi$, where (3.29) is multiplied by -1/2, and with $e_2 > 1$.

From the form of (3.52), we have

Lemma 3.41 *For* $0 < \mu \ll 1, 0 < r_2 \ll 1$, *(3.52) can be written as*

$$\tilde{G}_H = \frac{1}{2}\frac{\mu(e_2 - 1)}{r_2} - \frac{3}{2} + \tilde{\mathcal{O}}(\mu, r_2), \tag{3.53}$$

where $\tilde{\mathcal{O}} = \mathcal{O}_1((\mu r_2)^{1/2}) + \mathcal{O}_2(\mu) + \mathcal{O}_3(r_2)$, *where*

$$\mathcal{O}_1 = \mp r_2(\mu(e_2+1)r_2^{-1})^{\frac{1}{2}}, \quad \mathcal{O}_2 = \frac{\mu}{1+r_2} + \mu - \frac{1}{2}\mu^2 + \mathcal{O}(r_2),$$

$$\mathcal{O}_3 = -r_2(1-\mu),$$

and $\tilde{\mathcal{O}}$ *can be made arbitrarily small for* $0 < \mu \ll 1, 0 < r_2 \ll 1$. $\mathcal{O}(r_2) = -r_2 + \cdots$ *is a convergent binomial series in* r_2.

Note that the leading order term in (3.53) is positive since $e_2 > 1$. This will allow $\tilde{G}_H$ to be increased to desired values.

Lemma 3.41 is proven for perpendicular crossing of P_3 at $\bar{\mathbf{x}}(0)$ at a distance r_2 from P_2. Let β be the angle of $\dot{\bar{\mathbf{x}}}(0)$ with respect to the x_1-axis. Lemma 3.41 is proven for $\beta = \pm\pi/2$. However, the proof immediately generalizes for any value of $\beta \in [0, 2\pi]$, except $\beta = \pi$, to avoid collision of P_3 with P_1, provided $|\dot{\bar{\mathbf{x}}}| = |\pm\sqrt{\mu(e_2+1)/r_2} - r_2|$. This directly follows since the velocity dependence in $\tilde{G}$ is only on $|\dot{\bar{\mathbf{x}}}|$. Thus, we have the following lemma.

Lemma 3.42 *If*

$$|\dot{\bar{\mathbf{x}}}| = |\pm\sqrt{\mu(e_2+1)/r_2} - r_2|,$$

then Lemma 3.41 and hence (3.53) are true for any value of $\beta \in [0, 2\pi]$, $\beta \neq \pi$.

β is called the *crossing angle.* It is noted that as β varies in $[0, 2\pi], \beta \neq \pi$, then the angle ω of the location of the periapsis point of the parabola with the positive x_1-axis will vary between $[0, 2\pi], \omega \neq \pi$. ω is called the *argument of periapsis.*

It is recalled from section 3.1 that $0 < \mu \ll 1$ means μ is arbitrarily near to 0.

Now, notice that under the conditions of Lemma 3.42, the leading term of (3.53) *need not be arbitrarily small.* This fact is important for our results on matching $|C|$ with $\sqrt{2}$.

Equation (3.53) implies we would like to satisfy the equation

$$\frac{\mu(e_2 - 1)}{r_2} = 3 \mp \sqrt{2} - 2\tilde{\mathcal{O}}(r_2, \mu), \tag{3.54}$$

where $(\mu, r_2, e_2) \in T$,

$$T = \{0 < \mu \ll 1, 0 < r_2 \ll 1, e_2 \gtrsim 1\}, \tag{3.55}$$

and $\tilde{\mathcal{O}}(r_2, \mu) \ll 1$. Set

$$N = \frac{\mu(e_2 - 1)}{r_2}$$

and $\alpha^+ = 3 + 2\sqrt{2} - 2\tilde{\mathcal{O}}(r_2, \mu)$, $\alpha^- = 3 - 2\sqrt{2} - 2\tilde{\mathcal{O}}(r_2, \mu)$.

Lemma 3.43 *Assume $(\mu, r_2, e_2) \in T$. Let $\mu = \mu^*$ be fixed arbitrarily small, $\mu^* < \mu^{**} \ll 1$. Then, $N^* = \mu^*(e_2 - 1)r_2^{-1}, e_2 \gtrsim 1$, can achieve the value (3.54) by decreasing r_2 and/or $e_2 - 1$.*

Proof.

(i) If $\mu^*(e_2 - 1) \le r_2$, then $N^* \le 1$. If $N^* < \alpha^-$, then it can be increased to its desired value (3.54) by fixing e_2 and decreasing r_2 towards 0 to first achieve α^-, and then to achieve α^+ by decreasing r_2 further. If $1 > N^* > \alpha^-$ we can decrease it to the desired value by fixing r_2 and decreasing e_2 towards 1.

(ii) If $\mu^*(e_2 - 1) > r_2$, then $N^* > 1$. If $N^* > \alpha^+$, the desired values of N^* are obtained by fixing r_2 and decreasing e_2 towards 1 to first achieve α^+, and then α^- by further reduction. □

Since $\tilde{\mathcal{O}}$ is arbitrarily small on the set T then,

Lemma 3.44 *If $(\mu, r_2, e_2) \in T$, then by Lemma 3.43, r_2^* can be taken sufficiently close to zero, $0 < r_2 < \epsilon_1$, and/or e_2 sufficiently close to 1, $0 < e_2 - 1 < \epsilon_2$, for each fixed $\mu = \mu^*$ so that*

$$|\tilde{G}_H| = \sqrt{2} + \delta, \tag{3.56}$$

where $\delta > 0$ is any small number.

We make another extension to W and consider points where $\dot{r}_2 \ge 0$. This extension is added to the set W_H defined by (3.50), where σ^* is extended to the set

$$\tilde{\sigma}^* = \{r_2, \theta_2, \dot{r}_2, \dot{\theta}_2 | |\dot{r}_2| \ge 0\}.$$

This defines the set

$$\tilde{W}_H = W_H \cup \tilde{\sigma}^*;$$

since it is assumed that in $\tilde{W}_H$, $e_2 \gtrsim 1$, the points in $\tilde{W}_H$ are only slightly hyperbolic. The condition $e_2 \gtrsim 1$ is equivalent to $E_2 \gtrsim 0$. The assumption that $r_2 \gtrsim 0$ in the set T is consistent with the geometry of W since $r_2 \to 0$ for $e_2 \uparrow 1$, with $\dot{r}_2 = 0$. This is seen in (3.34), for $C \gtrsim C_1$.

Definition 3.45 *The* extended weak stability boundary *is given by the set*

$$\tilde{W} = W \cup \tilde{W}_H, \tag{3.57}$$

where on $\tilde{W}_H$, $e_2 \gtrsim 1$, $r_2 \gtrsim 0$, *and* $|\dot{r}_2| \geq 0, \mu \ll 1$. *On* W, $e_2 \in [0,1], r_2 > 0, \dot{r}_2 = 0, \mu \ll 1$.

It is noted that on $\tilde{W}$, both $\dot{\theta}_2, \dot{r}_2$ are bounded since $r_2 > 0$.

3.6 EXISTENCE OF A HYPERBOLIC NETWORK ON $\tilde{W}_H$

3.6.1 Introduction

We will prove that there exists a hyperbolic network on the extended weak stability boundary. This implies the existence of a complicated dynamics including permanent capture and unbounded oscillatory motion, among other types of motions.

In the last section we defined the extended weak stability boundary $\tilde{W}$ (see Definition 3.45),

$$\tilde{W} = W \cup \tilde{W}_H,$$

where on $W, 0 \leq e_2 \leq 1, r_2 > 0, \dot{r}_2 = 0$ and on $\tilde{W}_H, e_2 \gtrsim 1, r_2 \gtrsim 0, |\dot{r}_2| \geq 0$, and $\mu \ll 1$.

Thus, on $\tilde{W}_H, (r_2, e_2, \dot{r}_2, \mu) \in I_H = \{r_2, e_2, \dot{r}_2 | 0 < r_2 < \epsilon_1, 1 < e_2 < 1 + \epsilon_2, |\dot{r}_2| \geq 0, 0 < \mu < \epsilon_3, 0 \leq \epsilon_k \ll 1, k = 1,2,3\}$.

This implies that for $\dot{r}_2 = 0$, for each fixed $C > C_1$ and μ, $\tilde{W}_H$ is a small thin annulus (punctured disc, since $r_2 \gtrsim 0$) as follows from (3.33), centered about P_2. On this set, points have a very slight hyperbolic two-body energy with respect to P_2; that is, $E_2 \gtrsim 0$. For $\dot{r}_2 > 0$ the annulus has thickness in the $\dot{r}_2$ dimension.

It is noted from subsection 3.5.1 that the scaling of $\hat{J}$ by $-1/2$, implies that on $\tilde{W}, C > C_1$, where C is scaled from the value used prior to subsection 3.5.1 by $-1/2$ and the same symbol C is used. C_1 is $-1/2$ times the value (3.31).

The Jacobi constant C is fixed, and $C \in Q_1$. Q_1 is defined in Definition 3.23, where now $C > C_1$ due to the scaling of $\tilde{J}$. The set Q_1 is modified in the natural way.

If a solution of the restricted problem is on the set $\tilde{W}_H$ at a time t, then since $e_2 \gtrsim 1$, it is very close to ballistic capture, and by Definition 3.17 is called *pseudoballistic capture*

Now, although $\tilde{W}_H$ is a small annular set and near to $P_2 \equiv M$ in physical space, it turns out to intersect a hyperbolic network Λ formed from a transverse homoclinic point $\mathbf{r}$ of a two-dimensional area-preserving map ϕ. A set of this type is shown in Figure 2.8. The orbits of the restricted problem passing through the points of Λ are close to parabolic orbits. The map ϕ represents the return map of these orbits to $\tilde{W}_H \cap \Lambda$. The complexity of ϕ on Λ then yields a very complicated motion, containing permanent capture orbits and other types. This result implies, among other things, that there exists points of pseudoballistic capture which imply permanent capture. It also proves that there exists a hyperbolic network on $\tilde{W}_H$, and hence on $\tilde{W}$. Thus there exist points on $\tilde{W}$ which give rise to chaotic dynamics.

The existence of a hyperbolic network Λ in general near P_2 was proven by Xia [228]. His results and proof are given in subsection 3.6.5. We prove that Λ intersects $\tilde{W}_H$ in subsection 3.6.6. This is stated in Theorem 3.58.

Theorem 3.58 relates ballistic capture to permanent capture and, more generally, the extended weak stability boundary to chaotic dynamics. Applications and implications are discussed.

To describe these results, it will be necessary to describe the dynamics on a hyperbolic network associated to a transverse homoclinic point. Initially looking at a hyperbolic tangle as we saw in chapter 2 it would seem impossible to conveniently describe the dynamics of a map in this type of set. However, there is a convenient way to do this.

3.6.2 Preliminary Definitions and Theorems

To keep track of the iterates of a two-dimensional map ϕ on a hyperbolic network Λ, in general, we discuss a map called a *Bernoulli shift*. At first it won't be obvious that this has anything to do with ϕ or Λ, but a basic theorem by Smale relates them. Let

$$s = (\ldots, s_{-2}, s_{-1}, s_0; s_1, s_2, \ldots) \tag{3.58}$$

be a sequence of elements s_k, $s_k \in Q$. Q is the set of symbols being used. Q could be infinite or finite. S is the space of bi-infinite sequences with infinitely many elements s_k.

For example, the s_k could be integers, i.e., $Q \equiv \mathbb{Z}$. In this case there are infinitely many symbols. On the other hand, Q could be a set of two elements, say, $Q = [0,1]$. Thus, s is a sequence of 0 and 1 only: a binary sequence. We can associate to any binary sequence two numbers, a, b,

$$a = \sum_{k=0}^{-\infty} s_k 2^{k-1}, \qquad b = \sum_{k=1}^{\infty} s_k 2^{-k}. \tag{3.59}$$

This is a mapping

$$\mathbf{g} : S \to \square, \qquad \square = \{0 \leq a \leq 1, 0 \leq b \leq 1\}.$$

On the space S of elements (3.58), a neighborhood basis can be defined using

$$s^* = (\ldots, s^*_{-1}, s^*_0; s^*_1, s^*_2, \ldots\}.$$

The sets $N_j = \{s \in S | s_k = s^*_k, |k| < j\}, j = 1, 2, \ldots,$ form the basis. This defines a topology in S, thereby making S into a topological space. The map

$$(\phi(s))_k = s_{k-1}$$

is called a *shift map* on S. ϕ is a homeomorphism. When Q is finite, ϕ is called a *subshift of finite type*. A sequence s is unbounded if $\sup\{s_i, i \in \mathbb{Z}\} = \infty$. Also, as is described in [175], sequences can be defined which terminate or start in $\pm\infty$. In that case Q has infinitely many symbols.

A measure m can be defined on S making S into a measure space [175]. ϕ_k preserves m and is called a *Bernoulli shift*.

Returning to the example (3.59), the map $\boldsymbol{\rho} = \mathbf{g}\phi\mathbf{g}^{-1} : \square \to \square$ is measure preserving and given by

$$a_1 = 2a - [2a], \qquad b_1 = \frac{1}{2}(b + [2a]), \tag{3.60}$$

where $[a]$ is the largest integer $\leq a$. This transformation stretches rectangles by a factor of 2 in the horizontal direction and contracts by a factor of 1/2 in the vertical direction, and is called the *baker transformation* [175, 16, 68]. This terminology came about since the stretching and contracting of the rectangles is analogous to a baker manipulating dough to make bread.

The two vertical rectangles $R_1 = \{0 \leq a < 1/2\}, R_2 = \{1/2 \leq a \leq 1\}$ in the a,b-plane, where $b \in [0,1]$ are mapped by $\boldsymbol{\rho}$ into the two different horizontal rectangles $R'_1 = \{0 \leq b \leq 1/2\}, R'_2 = \{1/2 \leq b \leq 1\}$, where $a \in [0,1]$. They are obtained by contracting R_1, R_2 by a factor of 1/2, stretching them by a factor of 2, and laying one on top of the other to obtain R'_1, R'_2.

This contraction and stretching is analogous to the contraction and expansion in the neighborhood of a hyperbolic point. The baker transformation already shows that a shift with only two different symbols on bi-infinite sequences is related to hyperbolic contraction and expansion. This is an interesting fact. If the number of symbols increases from 2, then the map $\mathbf{g}$ can become much more complicated than $\boldsymbol{\rho}$ and can be made to be a more general contraction and expansion map. This type of result is expressed in the fundamental theorem by Smale [89, 175].

Theorem 3.46 (Smale-Birkhoff Theorem) *Let* $\mathbf{p} \in \mathbf{R^n}$ *be a hyperbolic fixed point for a diffeomorphism* $\mathbf{f} : \mathbf{R^n} \to \mathbf{R^n}$*, and assume that there exists a point* $\mathbf{r} \neq \mathbf{p}$ *where there is a transversal intersection of* $\mathbf{W^s(p)}, \mathbf{W^u(p)}$*. Then there exists a hyperbolic invariant set* Λ *(see Definition 3.47) on which* $\mathbf{f}$ *is topologically equivalent to a subshift of finite type.*

Definition 3.47 Λ *is a hyperbolic invariant set for* $\mathbf{f}$ *if it has a hyperbolic structure. This is a direct sum on the tangent space of* $\mathbf{R^n}$*:*

$$T_\Lambda \mathbf{R^n} = \mathbf{E^u_\Lambda} \oplus \mathbf{E^s_\Lambda},$$

where $\mathbf{E^u_\Lambda}, \mathbf{E^s_\Lambda}$ *are subspaces and where there exist constants* $\rho > 0, 0 < \sigma < 1$ *such that*

(i) *if* $\mathbf{v} \in \mathbf{E^u_p}, |D\mathbf{f^{-n}}(\mathbf{p})\mathbf{v}| \leq \rho\sigma^n|\mathbf{v}|$;

(ii) *if* $\mathbf{v} \in \mathbf{E^s_p}, |D\mathbf{f^n}(\mathbf{p})\mathbf{v}| \leq \rho\sigma^n|\mathbf{v}|$;

and $D\mathbf{f}$ *is the Jacobian matrix,* $D\mathbf{f} = (\partial f_i/\partial x_j), i, j = 1, \ldots, n, \mathbf{p} \in \Lambda$.

The hyperbolic network we discussed in chapter 2 for Figure 2.8 is an example of a hyperbolic invariant set. It is referred to as a *homoclinic tangle*, or equivalently a *hyperbolic tangle*. In the limit, each point of the network has an expanding and contracting direction. An important example of a hyperbolic invariant set Λ is given by the *Smale horseshoe*. This is a hyperbolic invariant limit set of a two-dimensional map $\mathbf{f}$ which expands, contracts, and folds a rectangle at each iteration; it is called a Smale horseshoe map. $\mathbf{f}$ maps the unit square into $\mathbf{R^2}$. By repeated applications of $\mathbf{f}$, the initial unit square is transformed into a set of thin horizontal and vertical strips, containing expanding and contracting directions. As more and more iterates of $\mathbf{f}$ are made, the strips get thinner and thinner, and in the limit a set of points is obtained as accumulation points of the intersections of the horizontal and vertical strips. Each point of the mesh has an expanding and contracting direction for the map. The invariant set Λ is a Cantor set, and it contains a dense orbit. It is topologically equivalent to a shift on two symbols. The Smale horseshoe is important since it appears whenever transverse homoclinic orbits for a flow occur. For an excellent discussion of the Smale horseshoe, hyperbolic invariant sets, and transverse homoclinic orbits, see [89, 175].

Theorem 3.46 is proven in [89] and in [175] for the case when $\mathbf{f}$ is area-preserving, with infinitely many symbols and for $n = 2$. The proof of this theorem is accomplished by looking for a Smale horseshoe–type set for an iterate of $\mathbf{f}$. For our applications, we require the version that Moser proves. We will state a version of Theorem 3.46 for this case below.

Theorem 3.46 is a powerful theorem with many applications. The condition of transversality of $\mathbf{W^s(p)}, \mathbf{W^u(p)}$ is often difficult to verify.

We can apply this theorem in the case of the existence of a transversal homoclinic orbit $\phi(t)$ *to a periodic solution* $\psi(t)$ *for a flow.* This situation is studied below for two examples. The reason the theorem can be applied in this case is because on a suitable surface of section to the transversal orbit, a map can be defined having the periodic orbit as a hyperbolic fixed point. This is accomplished by viewing the periodic orbit in the extended phase space where time t is a coordinate. If the periodic orbit has period T, then a surface of section $\sum_t$ to the flow is made for each value of $t = t_0 + kT, k = \pm1, \pm2, \ldots$, where t_0 is a given value of t. On this section, viewed as a snapshot in time, a map $\boldsymbol{\sigma}$ is defined with transversal invariant manifolds $\mathbf{W^u}_t, \mathbf{W^s}_t$ as slices of the manifolds $\mathbf{W^u}, \mathbf{W^s}$ of the flow on which the transversal homoclinic orbits exist. The section $\sum_t$ through the homoclinic orbit yields a homoclinic point r on the section. Since the orbit $\psi(t)$ has period T, $\mathbf{W^u}_t, \mathbf{W^s}_t$ are all identified at $t = t_0 + kT, k = \pm1, \pm2, \ldots$. This situation is shown in Figure 3.18.

Thus, transversality of the manifolds of the homoclinic orbit implies transversality of the manifolds on the section for the map. Therefore Theorem 3.46 can be applied. In applications the Melnikov method we describe later can be applied to verify that there exist transversal homoclinic orbits [89].

This situation shown in Figure 3.18, we refer to as the *breaking of a homoclinic loop associated to the hyperbolic point on* $\sum_t$. This occurs in two examples below in sections 3.5 and 3.6. The periodic orbit will have period $T = 2\pi$ in both cases and is a hyperbolic point in a time dependent system of two first order differential equations periodically depending on time with period T. In suitable coordinates x, y the hyperbolic point corresponds to $x = y = 0$. In extended phase space (x, y, t) the periodic orbit is just a line identified every 2π. The hyperbolic point will be shown to correspond to the set of parabolic orbits at infinity for the Sitnikov problem and the restricted problem. The second case considered in subsection 3.6.5 follows the situation in Figure 3.18. The notation $\sum_t$ used in subsection 3.6.5 will be used here as well. It will represent a *fixed time section* to the flow in extended phase space. When the time of the section is taken every integer multiple of the period 2π, then all the sections are identified as in Figure 3.18.

In the case where there are no transverse homoclinic orbits for the flow, then on each fixed time section there are no transverse homoclinic points for $\boldsymbol{\sigma}$. Then on $\sum_t$ there is a homoclinic loop as shown in Figure 2.7 on each section $\sum_t$. In this case, instead of Figure 3.18, we have Figure 2.13. The stable and unstable manifolds coincide, both for the flow in extended phase space and for the map on $\sum_t$. Since the manifolds coincide, then we obtain a topological cylinder as shown in Figure 2.13. For reference we call this a *homoclinic tube*, and when a transverse homoclinic orbit is proven to exist and Figure 3.18 results, we say that the *homoclinic tube breaks*.

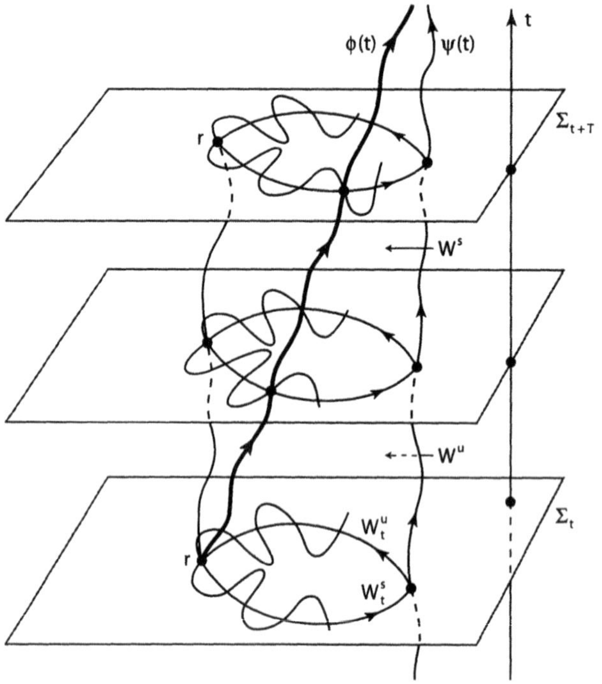

Figure 3.18 Time sections $\sum_t$ to periodic orbit ψ having a transverse homoclinic orbit ϕ. A map σ is defined on $\sum_t$ with transverse homoclinic point r. The sections are all identified.

3.6.3 Examples of Bi-Infinite Sequences and Associated Motions

We briefly show how bi-infinite sequences describe the possible dynamics of motions in the case of Sitnikov's problem given by (3.1), [175]. This general approach is referred to as *symbolic dynamics.* The radius term $r(t)$ in (3.1) is given by (1.47), where $a = 1/2, h = -1/2$. It is assumed that the eccentricity $e \ll 1$. In this case, it is verified that

$$r(t) = \frac{1}{2}(1 - e\cos t) + \mathcal{O}(e^2).$$

For $e = 0$ the Sitnikov problem is integrable [36]. The condition $e \ll 1$ is used to make various estimates from knowledge of the solutions for $e = 0$. In this way, e serves the purpose of being a small parameter as μ is in the restricted problem.

We recall the coordinate system Q_1, Q_2, Q_3 used to describe Sitnikov's problem, where P_3 was constrained to lie on the Q_3-axis.

Consider a solution $Q_3(t)$ to (3.1) for P_3. Assume $Q_3(t)$ has infinitely many zeros with the plane $Q_3 = 0$. That is, it oscillates up and down the Q_3-axis, passing through $Q_3 = 0$ infinitely often. Assume that $Q_3(t) = 0$ for a sequence $t = t_k, k = 0, \pm 1, \pm 2, \ldots$. These are ordered in magnitude,

$$t_k < t_{k+1},$$

$Q_3(t_k) = 0$. Set

$$s_k = \left[\frac{t_{k+1} - t_k}{2\pi} \right], \tag{3.61}$$

where the notation [ ] was introduced earlier. Thus, s_k gives a measure of the number of complete revolutions the primaries P_1, P_2 make (since they have period 2π) in the time it takes P_3 to make two passes through $Q_3 = 0$. With (3.61) a bi-infinite sequence (3.58) can be defined. Thus, every solution can be associated to such a sequence. Let's see what different sequences say. If the sequence is unbounded, then successive t_k become unbounded. This implies that the solution takes so long to come back to $Q_3 = 0$, it in fact is becoming unbounded, but it has infinitely many zeros. This is an unbounded oscillatory solution. For a periodic orbit a sequence is obtained which is periodic. For example, $s = \{\ldots, 1, 2, 1, 2, 1, 2, \ldots\}$

A permanent capture orbit which comes in from $|Q_3| = \infty$ corresponds to a sequence terminating on the left with ∞, and then performs infinitely many bounded oscillations. This implies

$$s = (\infty, s_k, s_{k+1}, \ldots),$$

$|s_k| < \infty$. If it just terminates on the right with ∞, then the orbit escapes to ∞ after performing infinitely many oscillations. Sequences terminating on both sides with ∞ correspond to orbits which come in from ∞, perform a finite number of oscillations, then escape. These are temporary capture orbits (Definition 3.9).

Now, if a two-dimensional map can be constructed for the flow of the Sitnikov problem on a surface of section, say $Q_3 = 0$, and this map can be proven to be topologically equivalent to a subshift on the above bi-infinite sequence space of *infinitely many symbols*, then Theorem 3.46, if extended to the case of infinitely many symbols, would then imply that all the above motions can be prescribed *independently*. That is, if a given bi-infinite sequence is prescribed, then a motion will exist that performs it. This is not surprising given the complexity of the hyperbolic invariant set. This type of result was proven by Alekseev [6] using the hyperbolic structure. However, his proof is unwieldy. A more direct and simplified presentation is by Moser [175]. He proves the following theorem.

Theorem 3.48 (Alekseev, Moser) *Given a sufficiently small eccentricity $e > 0$, there exists an integer $m = m(e)$ such that any sequence s, with $s_k \geq m$, corresponds to a solution of Sitnikov's problem. (3.1)*

The proof by Moser is an important contribution to celestial mechanics. His method has been used in other problems, as we will see in later subsections.

3.6.4 Brief Outline of Moser's Proof

This outline is very briefly described, and the details are in [175]. Moser's proof comprises an entire book, and we only give key steps.

1. The first step is to define a return map ϕ to the section $Q_3 = 0$. Solutions are prescribed on $Q_3 = 0$ with initial conditions $\dot{Q}_3(t_0) = \dot{Q}_3^\circ, Q_3(t_0) = 0$. We set $v_0 = |\dot{Q}_3^\circ|$. Because (3.1) is periodic in t, t_0, v_0 are polar coordinates on $Q_3 = 0$. t_0 is the angular variable, and v_0 is thought of as the radial variable. Assuming the solution returns to $Q_3 = 0$ at a time $t_1 > t_0$, this defines a map
$$\phi : (v_0, t_0) \to (v_1, t_1). \tag{3.62}$$

2. On the plane $Q_3 = 0$ there exists a closed curve γ, on whose interior D_0 ϕ is defined. The curve γ corresponds to parabolic orbits. The points outside of D_0 are escape solutions.

3. To better understand this map, a transformation due to McGehee [150] is used. It is given by
$$Q_3 = \frac{2}{x^2}, \qquad \dot{Q}_3 = -y, \qquad dt = 4x^{-3}ds, \tag{3.63}$$
$0 < x < \infty$. $x = 0$ corresponds to $Q_3 = \infty$. The point $x = 0, y = 0$ represents an equilibrium point of the transformed differential equations. They are of the form $\dot{x} = f(x, y, t), \dot{y} = g(x, y, t)$, where f, g are smooth in all variables and periodic of period 2π in t. Since these equations are periodic in t, the point $x = y = 0$ is actually a periodic orbit: $\mathbf{0} \times \{t\}, t \in \mathbf{R^1} \mod 2\pi$ in the extended phase space. We call this the *periodic orbit at* ∞. $x = y = 0$ is a degenerate hyperbolic point, with stable and unstable manifolds $\mathbf{W^u(0)}, \mathbf{W^s(0)}$.

   The manifolds are C^∞ functions $\Psi(y, t)$ and can be written, respectively, as $x = \Psi(y, t), x = \Psi(-y, -t)$, where Ψ has period 2π in t. In the extended phase space (x, y, t), they are topologically equivalent to cylinders as t varies from 0 to 2π.

On the stable manifold solutions spiral asymptotically to the periodic orbit as time approaches infinity, and the same is true for the unstable manifold as time approaches minus infinity. These manifolds are seen in Figure 3.19 for $e \neq 0$. It is remarked that the fixed point $(0,0)$ is degenerate since the Jacobian matrix at this point for the flow has eigenvalues equal to 1. Thus, the stable manifold theorem discussed in Section 2.3 cannot be used. McGehee [150] solved this problem by proving a stable manifold theorem in this situation.

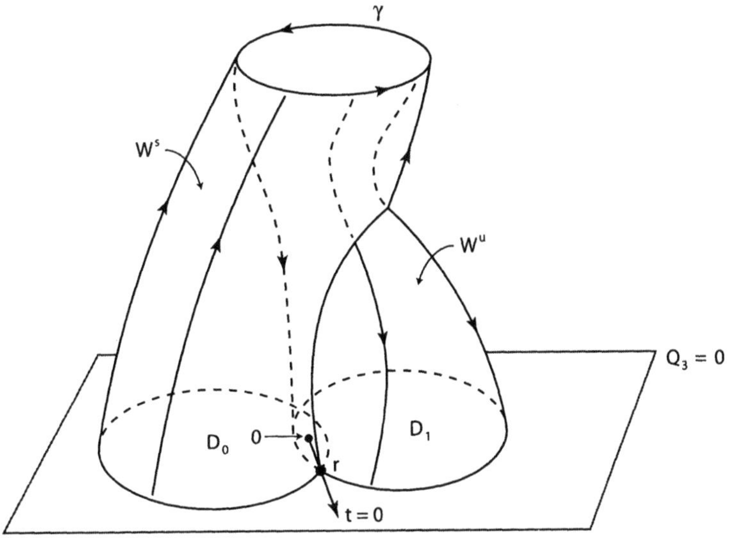

Figure 3.19 Intersection of $\mathbf{W^u(0)}, \mathbf{W^s(0)}$ on $Q_3 = 0$.

4. The invariant manifolds can be used to define a local coordinate system x, y near $x = y = 0$, corresponding to the parabolic solutions. The manifolds $\mathbf{W^u(0)}, \mathbf{W^s(0)}$ to $x = y = 0$ are the x, y-axis, respectively. These coordinates are obtained from the coordinates x, y by a smooth transformation $\tilde{x} = u(x, y, t), \tilde{y} = v(x, y, t)$, which are periodic of period 2π in t. For notational purposes, we relabel $\tilde{x}, \tilde{y}$ as x, y for convenience. A well-defined region near $x = y = 0$ is defined for orbits which are elliptic, parabolic, or hyperbolic. This is seen in Figure 3.20. The orbits are considered in the upper right quadrant. It is noted that although we have defined hyperbolic and parabolic orbits for the elliptic problem in section 3.1, we did not define elliptic orbits for this problem. Here, and for the remainder of this book, we define elliptic orbits as those orbits which are not hyperbolic or parabolic. That is, they never reach infinity. This definition of elliptic orbits is more general than the definition given in the two-body problem, where having negative Kepler energy with respect to the primary P_1 is required.

Here the Kepler energy could be positive for elliptic orbits in the more general setting.

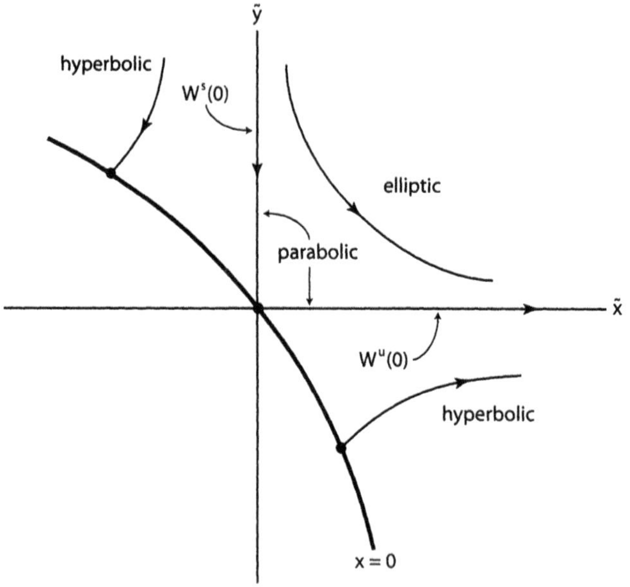

Figure 3.20 Region near infinity in McGehee coordinates.

5. A map $\psi : (x_0, y_0, t_0) \to (x_1, y_1, t_1)$ is defined near $x = y = 0$, which is obtained from ϕ given by (3.62). ψ is used to prove the existence of transverse homoclinic points and a hyperbolic invariant set for ϕ from a version of Theorem 3.46. The hyperbolic invariant set lies near the curve bounding D_0. ϕ maps thin annuli $S_0(\delta) : 0 < x < \delta, y = a$, where t is considered to be the angular variable, to annuli $S_1(\delta) : x = a, 0 < y < \delta,\ \ 0 < \delta \le a < 1$. $\psi : S_0 \to S_1$. See Figure 3.21. Following orbits from S_0 back to their first intersections on $Q_3 = 0$ for decreasing t defines the map $\phi_- : S_0 \to D_0$. Likewise, following orbits from S_1 forward to their first intersection on $Q_3 = 0$ defines $\phi_+ : S_1 \to D_1$. Thus,

$$\phi = \phi_+ \psi \phi_-^{-1} : \ D_0 \to D_1. \tag{3.64}$$

Remark. Each of the two-dimensional invariant manifolds $\mathbf{W^s(0)}, \mathbf{W^u(0)}$ for the flow in extended phase space intersect $Q_3 = 0$ on $Q_3 = 0$ in a simple closed curve with interiors D_0, D_1, respectively. These two closed curves intersect in a point $\mathbf{r}$ (see Figure 3.19). This is a transversal homoclinic

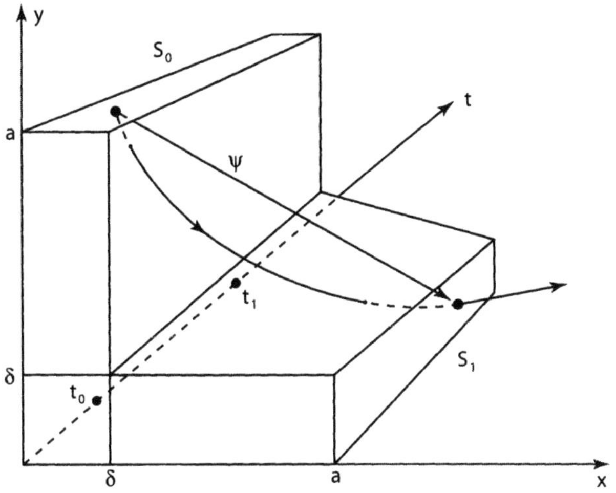

Figure 3.21 Map ψ near infinity.

point of a homoclinic orbit and is an analog for a homoclinic point of the map ϕ.

Let S be the space of bi-infinite sequences containing infinitely many symbols, which can terminate with ∞, and let $\boldsymbol{\sigma}$ be a shift map on S.

Theorem 3.49 *The mapping ϕ in D_0 has an invariant hyperbolic set $\Lambda \subset D_0$ which is homeomorphic to S, and ϕ on Λ is topologically conjugate to $\boldsymbol{\sigma}$ on S. That is, let $\mathbf{h}$ be the homeomorphism, $\mathbf{h} : S \to \Lambda$; then $\phi = \mathbf{h}\boldsymbol{\sigma}\mathbf{h}^{-1}$.*

This represents the main theorem of [175], from which Theorem 3.48 follows. The map ϕ on $Q_3 = 0$ is proven to have a transverse homoclinic point and has the typical appearance, as is seen in Figure 3.22

The existence of the invariant set Λ implies the nonexistence of a real analytic integral for (3.1), except the energy,

Theorem 3.50 (Moser) *There does not exist a real analytic integral for the Sitnikov problem.*

The idea of the proof is that if there exists a constant function f on Λ, then the directional derivatives of f at each point of Λ are zero along the hyperbolic directions at each point. Since Λ is dense and the hyperbolic directions form an infinitesimal grid on Λ, then the derivatives in all directions are zero. This forces $f \equiv 0$.

The proof of Theorem 3.49 follows from a version of the Smale-Birkhoff theorem. This is Theorem 3.51 stated below. This theorem is proven not for ϕ, but for the *transversal map* $\tilde{\phi}$ of ϕ. The transversal map plays a key role in Moser's proof.

We define the transversal map. More generally, let $\mathbf{p}$ be a hyperbolic point with manifolds $\mathbf{W^u(p)}, \mathbf{W^s(p)}$ for a diffeomorphism ϕ. Looking at Figure 3.22, a small quadrilateral R is placed near the homoclinic point $\mathbf{r}$. Two sides of R are $\mathbf{W^u(p)}, \mathbf{W^s(p)}$, and the other sides are straight lines parallel to the manifolds. For a point $\mathbf{q}$, let $k(\mathbf{q})$ be the smallest positive integer such that $\phi^k(\mathbf{q}) \in R$, if it exists. Let the set of $\mathbf{q} \in R$ for which such a $k > 0$ exists be denoted by D, and

$$\tilde{\phi}(\mathbf{q}) = \phi^k(\mathbf{q}) \tag{3.65}$$

for $\mathbf{q} \in D$. $\tilde{\phi}$ is called the *transversal map* of ϕ for R.

The version of the Smale-Birkhoff Theorem proven by Moser used to prove Theorem 3.49 is given by the following theorem.

Theorem 3.51 *If a C^∞-diffeomorphism ϕ has a homoclinic point $\mathbf{r}$ at which the invariant manifolds $\mathbf{W^u(p)}, \mathbf{W^s(p)}$ associated to a hyperbolic fixed point $\mathbf{p}$ intersect transversally, then in any neighborhood of $\mathbf{r}$, the transversal map $\tilde{\phi}$ of a quadrilateral has an invariant subset Λ homeomorphic to the set S of bi-infinite sequence on infinitely many symbols via a homeomorphism $\mathbf{h} : S \to \Lambda$ such that*

$$\tilde{\phi} = \mathbf{h}\sigma\mathbf{h}^{-1},$$

where σ is a shift on S.

This differs from the Smale-Birkhoff theorem in the main respect that the bi-infinite sequences have infinitely many symbols. It is also for the case where $n = 2$ and where the map $\tilde{\phi}$ is area-preserving. Theorem 3.51 yields the result that the set of homoclinic points belonging to $\mathbf{p}$ are dense in Λ.

We now relate the general transversal map to our maps ϕ and ψ for the Sitnikov problem, so that Theorem 3.49 can be deduced from Theorem 3.51. We describe this by referring to Figure 3.22.

In Figure 3.22, the strips S_0, S_1 shown in Figure 3.21 are projected onto the x, y-plane, on which ψ is defined; $\psi : S_0 \to S_1$, is discussed in step 5 earlier in this subsection. A homoclinic point $\mathbf{r}$ is shown and near it a small quadrilateral R, with images on S_0, S_1 of $\phi^\ell(\mathbf{r}), \phi^{-m}(\mathbf{r})$, respectively, for integers ℓ, m. $R_0 = \phi^\ell(R), R_1 = \phi^{-m}(R)$. $R_0 \subset S_0, R_1 \subset S_1$.

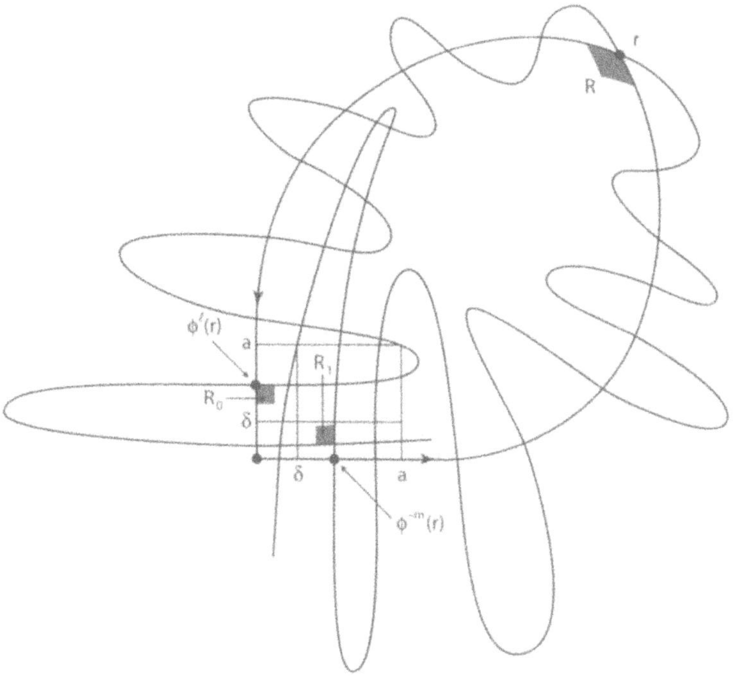

Figure 3.22 Construction of the transversal map, $\tilde{\phi}$.

Set $Q = \{0 \le x \le a, 0 \le y \le a\}$. With this setup we can define ψ naturally as follows: Let $\mathbf{q} \in R_0$ then there exists a smallest $k > 0$ such that $\phi^k(q) \in R_1$ and $\phi(\mathbf{q}), \phi^2(\mathbf{q}), \ldots, \phi^{k-1}(\mathbf{q}) \in Q$. Then

$$\psi(\mathbf{q}) = \phi^k(\mathbf{q}) \tag{3.66}$$

defines ψ. The domain of such $\mathbf{q} \in R_0$ is called D. In terms of $\tilde{\phi}$,

$$\tilde{\phi} = \phi^m \psi \phi^\ell.$$

This gives a summary of the key general steps in Moser's proof. The details and estimates are not included for brevity. The methodology of Moser's proof is applied to the planar restricted problem in the next subsection.

3.6.5 Transversal Homoclinic Orbits in the Restricted Problem

A theorem analogous to Theorem 3.49 is proven for the planar circular restricted problem by Xia [228], which we describe. This result will be used to prove the main result of the next section on the existence of a hyperbolic invariant set on $\tilde{W}_H$ and hence for $\tilde{W}$. This yields insights into ballistic capture and proves that it is a chaotic process.

As noted in section 3.1, a theorem analogous to Theorem 3.49 for the restricted problem was proven earlier by Llibre and Simó [137, 138]. They considered the case of C sufficiently large and μ sufficiently small. Xia extended these results for all but a finite number of μ in $(0, 1)$ and to the case of C near $\pm\sqrt{2}$, where μ is small. The situation of interest here is μ small and $|C|$ near $\sqrt{2}$.

Llibre, Simó, and Xia generally followed the procedure of [175] as outlined with various modifications. The idea is to define a two-dimensional map near parabolic orbits and insure that the map is well defined by using the transformation by McGehee; see (3.63). In McGehee coordinates x, y, the parabolic orbits at infinity correspond to $x = y = 0$. Next, the flow is studied near $x = y = 0$, a hyperbolic fixed point, which is a periodic orbit γ since the system is periodically time dependent. Then show that the invariant manifolds of the hyperbolic periodic orbit in extended phase space have transverse intersection by the existence of a transverse homoclinic orbit, thus yielding an invariant hyperbolic set by the Smale-Birkhoff theorem by a reduction to a map, as we described in subsection 3.6.2. Unlike the approach in [175] Llibre, Simó, and Xia used the Melnikov method to prove the transversality of the manifolds from infinity by proving the existence of transverse homoclinic orbits.

Xia's proof is relatively simpler, and it follows the procedure outlined in subsection 3.6.2 following the idea of the construction of Figure 3.18. Direct appeal is made to the Smale-Birkhoff theorem. Moser did not do this, as he actually proved a version of the Smale-Birkhoff theorem, which is involved. Also, Xia's proof is further simplified by using the theorem of Easton and Robinson in subsection 3.5.4 on the existence of parabolic orbits in the planar three-body problem, Theorem 3.38.

Xia's approach followed these steps: 1. Transform to McGehee coordinates 2. Set $\mu = 0$ 3. Solve the two-body problem and using Theorem 3.38 to find a homoclinic loop for fixed time sections 4. For $\mu \neq 0$, show that the homoclinic tube breaks by using the Melnikov Method. 5. Apply the Smale-Birkhoff theorem.

Steps 1, 2, 3 closely follow Easton and Robinson's construction for the planar three-body problem of which this is a special case, so one can compare the construction in [228] in the McGehee coordinates with [76, 78, 195]. Step 4 uses the Melnikov method as presented in [89]. The Melnikov calculation carried out in detail in [228]. The Smale-Birkhoff theorem is applied in a straightforward manner and closely follows [175].

The proof is carried out in coordinates which do not readily lend themselves to an intuitive physical understanding of the dynamics. We will describe the dynamics and the maps from a more physical standpoint. This is necessary for the next subsection. We also expand on various aspects of Xia's proof and relate it to previous results.

It is remarked that all these results, including those in [175], can be viewed as applications of the Smale-Birkhoff theorem, thus showing the importance of this theorem. This theorem can be roughly stated as the fact that under suitable conditions, transversality of hyperbolic manifolds of maps implies chaos. Since this transversality can occur due to the breaking of a homoclinic loop, then the existence of a homoclinic loop in a system is a sign that chaos may exist. A homoclinic loop can be viewed as a smoking gun for existence of chaos.

We will outline the proof in [228], and the reader can find further details in that paper.

The restricted problem is considered in inertial barycentric coordinates (Q_1, Q_2), Equation (1.55), and the Jacobi integral given by (1.58). Equation (1.55) is time dependent and periodic of period 2π. This time dependence is analogous to the Sitnikov problem and motivates why inertial coordinates are used. A map analogous to ϕ in the previous subsection is defined, which we also label ϕ. We assume at $t = 0$, that P_3 is on the Q_1-axis. As shown below for a given value of C and at the distance $|Q_1(0)|$, $|\dot{\mathbf{Q}}(\mathbf{0})|$ can be uniquely adjusted so that P_3 is an ω-parabolic orbit, which by symmetry $(Q_1 \to Q_1, Q_2 \to -Q_2, t \to -t)$ is an α-parabolic orbit. If $|\dot{\mathbf{Q}}(0)|$ is not large enough, then P_3 will not escape to infinity. Assume P_3 intersects the Q_1-axis infinitely many times.

Let s_k be the number of revolutions of P_1, P_2 between successive times $t_k, k = \pm 1, \pm 2, \ldots$ of intersection of P_3 with the Q_1-axis, where $t_k < t_{k+1}$. s_k is defined as in (3.61). This defines as in the previous subsection a space S of bi-infinite sequences s of infinitely many symbols.

This definition of the bi-infinite sequences is not used in [228]. With this definition of S, the central result of [228] can be formulated as the following theorem.

Theorem 3.52 *For (i) μ sufficiently small and $|C| \gtrsim \sqrt{2}$, or (ii) $|C|$ sufficiently large and all but a finite set of $\mu \in (0,1)$, there exists an integer $m = m(\mu, C)$ such that any sequence $s \in S$ with $s_k \geq m$ corresponds to a solution of the restricted problem.*

This is analogous to Theorem 3.48. Unbounded sequences give rise to unbounded oscillating orbits, and as in the previous subsection, permanent capture is obtained by sequences terminating on either the right or left side with ∞.

Before proceeding to describe a proof of this theorem, we discuss here how to determine the set of parabolic orbits on the Q_1-axis on a fixed time section for $t = 0$

For $\mu = 0$, we have the two-body problem between P_3, P_1, where P_1 is at the origin of mass 1. Equation (3.37) implies $J = \frac{1}{2}|\dot{\mathbf{Q}}|^2 - |\mathbf{Q}|^{-1} - c = H - c$, where H is the Kepler energy and c is the angular momentum. Both H and c are constants of the motion, and at $t = 0$ and on the Q_1-axis, $c = Q_1\dot{Q}_2$. Also, $H = 0$ for the set of parabolic orbits. Given $Q_1(0)$, a family of parabolic orbits on the Q_1-axis at the distance $|Q_1(0)|$ is obtained where $|\dot{\mathbf{Q}}| = (2|Q_1|)^{-\frac{1}{2}}$ is determined from $H = 0$. This implies for each value of $Q_1(0)$ there is a circle of parabolic orbits satisfying $|\dot{\mathbf{Q}}| = \mathit{constant}$. This means that for any angle $\beta \in [0, 2\pi]$ a parabolic orbit will cross the Q_1-axis whose tangent vector makes the angle β with the the axis. Thus, the set of ω or α-parabolic orbits is two-dimensional, parameterized by $Q_1(0), \beta$.

For $\mu \neq 0$, c is not a constant, and neither is H. By Theorem 3.52 applied to the restricted problem as in Theorem 3.54 below, for μ sufficiently small, the set of parabolic orbits in the restricted problem lie close to the set in the two-body problem and also comprise a two-dimensional set. As we will see in this subsection, these parabolic orbits in general are associated to a hyperbolic invariant set and are unstable.

These orbits can also be described relative to $|\mathbf{Q}| = \infty$ by passing to McGehee coordinates, as we did in Sitnikov's problem. We choose

$$\mathbf{Q} = x^{-2}\mathbf{s}, \qquad \dot{\mathbf{Q}} = y\mathbf{s} + x^2\rho i\mathbf{s}, \qquad i^2 = -1;$$

$\mathbf{s} = (\cos\theta, \sin\theta)$ is an angular component, and $x^2\rho$ is the norm of the radial component. $i\mathbf{s}$ is a unit vector, in complex notation, normal to $\mathbf{s}$. ρ is the angular momentum of P_3. In these coordinates The Jacobi energy (3.37) becomes

$$\Phi = \frac{1}{2}y^2 + \frac{1}{2}x^4\rho^2 - \Omega - \rho = C. \tag{3.67}$$

A new angular variable is defined,

$$s = t - \theta, \tag{3.68}$$

where $\theta = \theta(t)$ and s is also a variable on the circle. It is assumed that $\theta(0) = 0$. This implies that $t = 0$ maps into $s = 0$. $\theta = 0$ corresponds to the Q_1-axis. This choice of variables is used in [76, 195]

Set $\mu = 0$. It is verified that (1.55) becomes for $\mu = 0$

$$\dot{x} = -\frac{1}{2}x^3y, \qquad \dot{y} = -x^4 + x^6\rho^2, \qquad \dot{s} = 1 - x^4\rho, \tag{3.69}$$

$\rho =$ constant. The first two equations are generally independent of s and can be explicitly solved. The constant energy function Φ of P_3 in this case is

$$\Phi = \frac{1}{2}y^2 + \frac{1}{2}x^4\rho^2 - x^2. \tag{3.70}$$

Level curves of Φ in the x, y coordinates for fixed C are shown in Figure 3.23 parameterized by ρ. The flow of the differential equations is also shown. This plot is in phase space x, y. $x > 0$ is where the actual motion occurs. This is slightly different from the Sitnikov problem studied by Moser, where $x > 0, y > 0$, as seen in Figure 3.20.

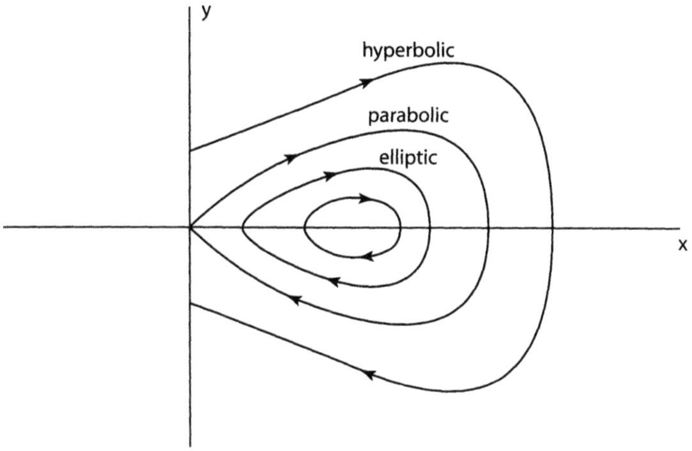

Figure 3.23 Kepler flow in McGehee coordinates.

For each C and ρ a single parabolic orbit is obtained intersecting the x-axis normally at $x = x_0$ in these coordinates. It approaches the point $x = y = 0$ as $t \to \pm\infty$. The specification of a parabolic orbit by C, ρ is analogous to the situation in our previous discussion, where in the coordinates $\mathbf{Q}, \dot{\mathbf{Q}}$, H, c specified a parabolic orbit.

When $C = \pm\sqrt{2}$, then $x_0 = 1$. The $-$ sign corresponds to direct parabolic motion, as we saw in the last section. $x = 1$ implies $|\mathbf{Q}| = 1$. Since we are taking initial conditions for the flow in the x, y coordinates at $s = 0$ where $\theta = 0$, then the initial position is on the Q_1-axis and for $C = \pm\sqrt{2}$. P_3 is at the location of P_2 for $\mu = 0$ where $r_2 = 0, r_{13} = 1$. We will be assuming later that $|C| \gtrsim \sqrt{2}$ and this implies that P_3 is at a distance $r_2 \gtrsim 0$ from P_2, where $r_{13} = 1 + r_2$.

Hyperbolic and elliptic orbits are shown in Figure 3.23. $x = y = 0$ is a singular point of the flow of (3.69) in x, y-space, where $\dot{x} = \dot{y} = 0$ $x = y = 0$ is a degenerate fixed point, and rescaling by a factor x^{-3}, implies it is a hyperbolic saddle. This same figure is shown in [78, 76, 195].

Constructing fixed time sections $\sum_s$ for the flow in Figure 3.23, we obtain a discrete time map defined on the homoclinic loop as pictured representing the motion of parabolic orbits. Figure 3.23 then represents a fixed time section $\sum_s$, and it is identified at $s \mod 2\pi$. The ω, α-parabolic orbits form the homoclinic loop, with one-dimensional manifolds $\mathbf{W^u(0)}, \mathbf{W^s(0)}$, which are identical. These time sections are equivalent to the situation shown in Figure 2.13, where the flow is viewed as taking place in the extended phase space x, y, s. We then obtain a homoclinic tube from the homoclinic loop in Figure 3.23, by considering the extended (x, y, s)-space, analogous to the situation shown in Figure 2.13. The idea of the proof is to show that the tube, or equivalently the homoclinic loop on fixed time sections, breaks for $\mu \neq 0$.

The homoclinic loop can explicitly be determined as

$$x(t) = \xi(t, C) = X_0(s, C),\ y(t) = \eta(t, C) = Y_0(s, C), \qquad (3.71)$$

where ξ is an even function of t and η is an odd function of t. This is analogous to the representation of the invariant manifolds mentioned in step 3 of Moser's proof. The loop crosses the x-axis perpendicularly for $x > 0$. We let

$$\mathcal{H} = \mathcal{H}(C)$$

denote this loop. Therefore, by our previous comments when $C = \pm\sqrt{2}$, the corresponding parabolic orbits cross the x-axis at the position of P_2. $|C| \gtrsim \sqrt{2}$ then implies at $t = 0, Q_1 \gtrsim 1$. We summarize some of these comments in the next lemma.

Lemma 3.53 *For $\mu = 0$, the one-dimensional homoclinic loop $\mathcal{H}$ passes through the position of P_2 with $m_2 = 0$ for $|C| = \sqrt{2}$ on the Q_1-axis, where $r_2 = 0$. For $|C| \gtrsim \sqrt{2}$, $\mathcal{H}$ satisfies $r_2 \gtrsim 0$ on the Q_1-axis, where P_3 is at the location where $Q_1 = 1 + r_2$.*

For $\mu \geq 0$, $x = y = 0$ represents a periodic orbit γ at ∞. This is the case since for $\mu \geq 0$ the differential equations for the restricted problem can be written as a system

$$x' = f_1(x, y, s, \mu, C), \qquad y' = f_2(x, y, s, \mu, C)$$

where f_1, f_2 are periodic functions of s of period 2π. Thus, as s varies over $[0, 2\pi]$ the point $(x, y) = (0, 0)$ is a periodic orbit. This is analogous to the situation in the Sitnikov problem where a similar system of differential equations is derived as we mentioned in step 3 of Moser's proof.

Theorem 3.38 in this case takes the following form.

Theorem 3.54 $\mathbf{W^u}(\gamma), \mathbf{W^s}(\gamma)$ *in extended phase space (x, y, t) for $\mu \neq 0$ and for a fixed value of the Jacobi constant are two-dimensional and are* *associated with a periodic orbit γ at $x = y = 0$, parameterized by $s \in [0, 2\pi]$. They exist for $x > 0$ and are real analytic. That is, the ω, α-parabolic orbits form real two-dimensional analytic submanifolds of the phase space for fixed Jacobi constant C.*

For $\mu = 0$ the parabolic orbits exist on $\mathcal{H}$ in (x, y)-space. In extended phase space they lie on the invariant manifolds $\mathbf{W^u} \equiv \mathbf{W^s}$ forming a homoclinic tube, as we discussed in subsection 3.6.2.

It is remarked that as P_3 is moving, it may collide with P_1 or P_2. However, such collision singularities can be regularized by the Levi-Civita transformation we studied in chapter 1. This same regularization would also regularize $\mathbf{W^u}$, $\mathbf{W^s}$, and in the regularized coordinates they would remain real analytic. However, this situation will not arise. C will be chosen so that collision will not occur. This is insured by choosing $|C| > \sqrt{2}$. This implies $|\mathbf{Q}| > 1$, and therefore P_3 will lie outside the orbit of P_2. Lemma 3.53 says that for $\mu = 0$, $\mathcal{H}$ lies slightly beyond P_2 on the Q_1-axis, at a distance $r_2 \gtrsim 0$, for $|C| \gtrsim \sqrt{2}$. This is also true for $\mu \gtrsim 0$ by the smoothness of $\mathbf{W^u}, \mathbf{W^s}$.

The main result in [228] is that for $\mu \neq 0$, $\mathbf{W^u}(\gamma), \mathbf{W^s}(\gamma)$ do not coincide and have transverse intersection for a set of μ, C.

Theorem 3.55 *For (i) μ sufficiently small and $|C| \gtrsim \sqrt{2}$, or (ii) $|C|$ sufficiently large and all but a finite set of $\mu \in (0, 1)$, $\mathbf{W^u}(\gamma), \mathbf{W^s}(\gamma)$ have transverse intersection due to the existence of a transverse homoclinic orbit.*

This is proved by the Melnikov method. The basic procedure is outlined here as it applies to our problem.

The proof is carried out on the Q_1-axis for $Q_1 > 1$ or equivalently for $|C| > \sqrt{2}$. The orbits start at this location at $t = 0$ corresponding to $s = 0$. The flow of $x' = f_1, y' = f_2$ for fixed C, μ is propagated from the time section $\sum_{s_0}$ located on the Q_1-axis. Orbits $\mathbf{x}_+(s, s_0, \mu)$ lying in $\mathbf{W^s}$ are considered where $\mathbf{x}_+(0, s_0, \mu) \in \sum_{s_0}$ and similarly orbits $\mathbf{x}_-(s, s_0, \mu)$ lying in $\mathbf{W^u}$ are considered where $\mathbf{x}_-(0, s_0, \mu) \in \sum_{s_0}$. $\mathbf{x}_+(s, s_0, \mu), \mathbf{x}_-(s, s_0, \mu)$ are *based on* $\sum_{s_0}$.

The Melnikov method is a procedure to measure the difference

$$\mathbf{d}(s_0) = \mathbf{x}_+(s_0, s_0, \mu) - \mathbf{x}_-(s_0, s_0, \mu).$$

This measures the slitting of the invariant manifolds at $s = s_0$ on $\sum_{s_0}$. The estimation of $\mathbf{d}$ is accomplished by first expanding

$$\mathbf{x}_+(s, s_0, \mu) = (X_0(s - s_0), Y_0(s - s_0)) + \mu \mathbf{v}_+(s, s_0) + \mathcal{O}(\mu^2),$$

$$\mathbf{x}_-(s, s_0, \mu) = (X_0(s - s_0), Y_0(s - s_0)) + \mu \mathbf{v}_-(s, s_0) + \mathcal{O}(\mu^2),$$

where for $\mathbf{x}_+$, $s \in [s_0, \infty)$, and for $\mathbf{x}_-$, $s \in (-\infty, s_0]$, $|C| > \sqrt{2}$. $\mathbf{v}_{+,-}(s, s_0) \in \mathbf{R}^2$ are determined by solving the first variational equation of $x' = f_1(x, y, s, \mu, C), y' = f_2(x, y, s, \mu, C)$ along the unperturbed orbit $X_0(s)$, $Y_0(s)$, based on $\sum_{s_0}$.

Consider the normal direction $\mathbf{n}_0$ to the surface $\Phi^{-1}(C)$, where Φ is given by (3.70) at the point $(x, y) = (X_0(0), Y_0(0))$. $\mathbf{n}_0 = (\Phi_x(X_0(0), Y_0(0)), \Phi_y(X_0(0), Y_0(0)))$ Let $d(s_0)$ be the projection of $\mathbf{d}(s_0)$ onto $\mathbf{n}$. $d(s_0)$ then approximates the separation of $\mathbf{W^u}, \mathbf{W^s}$ on the section $\sum_{s_0}$ at the point $(X_0(0), Y_0(0))$. The calculation of $d(s_0)$ is accomplished by estimating it with an integral, which is called the *Melnikov integral.* It is given by

$$M(s_0) = \int_{-\infty}^{\infty} \mu^{-1} \frac{d\Phi}{ds}(X_0(s - s_0), Y_0(s - s_0))ds$$

and

$$d(s_0)|\mathbf{n}_0| = \mu M(s_0) + \mathcal{O}(\mu^2).$$

Thus, $M(s_0)$ gives a measure of the separation of the manifolds. When $M(s_0) = 0$, then this can be proven to imply the manifolds intersect for μ sufficiently small. That is, $d(s_0) = 0$. However, this intersection may may not be transversal. To insure transversal intersection, it is necessary to prove that $M(s_0)$ has simple zeros. This represents the Melnikov method. We state this as a theorem [89].

Theorem 3.56 *If $M(s_0)$ has simple zeros and is independent of μ, then for $\mu > 0$ sufficiently small, $\mathbf{W^u}(\gamma), \mathbf{W^s}(\gamma)$ have transverse intersection.*

We need $M(s_0)$ independent of μ, since then the term $\mu M(s_0)$ dominates the expression for $d(s_0)$. This theorem implies a transverse homoclinic point on a time section, from our previous discussion, and then the Smale-Birkhoff theorem can be applied.

It is proven in [228] that $M(s_0)$ has a simple zero for $s_0 = \pi$. In order to show this it is necessary to assume that $|C| - \sqrt{2} \gtrsim 0$ for $\mu = 0$, which by continuity with respect to μ is valid for μ sufficiently small. The explicit estimation of the Melnikov integral in [228] is computed for $t = 0$ on the Q_1-axis. It is found that

$$M(s_0) = \int_{-\infty}^{\infty} X_0^4(s) \sin(s + \pi)(1 - r^{-3/2}(s, s_0))dt,$$

where $r(s, s_0) = 1 + 2X_0^2(s)\cos(s + s_0) + X_0^4(s)$. The integration variable is t, and s appears in the integrand. The relationship between s and t is given by

$$s = t - \theta(t) = t + \int_0^t \xi^4(t, C) C dt.$$

The integrand for $M(s_0)$ is odd, and the integration interval is symmetric. Thus,

$$M(s_0) = 0$$

for $s_0 = \pi$. To insure that this zero is a simple one, we must show that $M'(s_0)|_{s_0=\pi} \neq 0$. This follows by direct calculation of M', and it is checked that

$$M'(\pi) = \int_{-\infty}^{\infty} X_0^4(s) \cos(s+\pi)(1 - r^{-3/2}(s,\pi))dt$$
$$- \int_{-\infty}^{\infty} 3X_0^6 \sin^2(s+\pi) r^{-5/2}(s,\pi)dt.$$

Now, when $C = \pm\sqrt{2}$, then $r(s,\pi) = 0$ when $t = 0$, or equivalently when $s = 0$, where $\theta = 0$. We saw previously that when $C = \pm\sqrt{2}$, then parabolic orbits for $\mu = 0$ intersect the x-axis at $x = 1$. The function X_0 occurs for $\mu = 0$, and at $t = 0$, $X_0 = 1$. Therefore when $t = 0$,

$$r(0, \pi) = 2 + 2\cos(\pi) = 0.$$

This implies that $M'(\pi)$ is singular at $t = 0$, and as $|C| \to \sqrt{2}$, $|C| \gtrsim \sqrt{2}$, then $M'(\pi) \to \infty$. Thus for μ, $|C| - \sqrt{2}$ sufficiently small, Theorem 3.56 implies the existence of transverse homoclinic orbits.

This proves case (i) of Theorem 3.55. case (ii) of is proven for C sufficiently large and for all but a finite set of $\mu \in [0,1]$ by using, in part, Lagrangian intersection theory [223].

We now expand a bit on the domains of definition for the map ϕ we defined at the beginning of this subsection. This is done to give a more physical understanding to the map. Like the map ϕ in Moser's proof, it is defined on a two-dimensional set D_0. In our case, this set is defined on the positive Q_1-axis on the section $\sum_{t=0}$ for $|C| > \sqrt{2}$. The set D_0 can be obtained from the map ψ which is constructed in our case the same way as in step 5 in Moser's proof, near $x = y = 0$, as shown in Figure 3.21. ϕ is given by a map analogous to (3.64), as a function of ψ, where $\phi : D_0 \to D_1$. For a fixed value of C, D_0 is two-dimensional and can be described with the coordinates $Q_1(0), V(0) = |\dot{\mathbf{Q}}(\mathbf{0})|$.

It has a boundary ∂D_0 which is well defined, $\partial D_0 = \mathbf{W^s} \cap \sum_0 \cap \{Q_2 = 0\}$. This just represents the slice of $\mathbf{W^s}$ for $\theta = 0$. For each value of $Q_1(0)$ for $\theta = 0$, there is a critical value of $V(0) = V^*(0)$ such that the solution is ω-parabolic. This value is unique and is guaranteed by Theorem 3.54. Moreover, for

$$V(0) < V^*(0)$$

the motion is elliptic and does not reach infinity, and for

$$V(0) > V^*(0)$$

the orbit is hyperbolic. $V^*(0) > 0$ represents $\partial D_0 \equiv \mathbf{W}^s$ for this value of $Q_1(0)$ and θ. For $Q_1(0) \in [a,b], 0 < a < b$, ∂D_0 is a smooth simply connected curve $\alpha(a,b)$. This is a curve in Q_1, V-space lying above the Q_1-axis. This is seen to be the case for $\mu = 0$ from our earlier discussion in this subsection, where the norm of the velocity varies smoothly with the distance $|Q_1(0)|$, obtained from $H = 0$. A piece of the D_0 region on $\sum_0$ in Q_1, V-space is bounded by the two vertical lines $Q_1(0) = a, Q_1(0) = b$ and between $V(0) \in \alpha(a,b), V(0) = 0$.

The Smale-Birkhoff theorem extended to infinitely many symbols given by Theorem 3.51 can be applied in view of Theorem 3.55 on $\sum_t$, which implies a hyperbolic invariant set Λ for the two-dimensional map ϕ on $\sum_0$.

We obtain a theorem analogous to Theorem 3.49, which we state for the transversal map $\tilde{\phi}$, defined on a set D in the same way as in the previous subsection shown in Figure 3.22 for a small quadrilateral R near a homoclinic point $\mathbf{r}$ of ϕ. D is defined in the last subsection for the map ϕ, as (3.65). Analogous to Theorem 3.49 we have the following theorem.

Theorem 3.57 *For (i) μ sufficiently small and $|C| \gtrsim \sqrt{2}$, and (ii) $|C|$ sufficiently large and all but a finite set of $\mu \in (0,1)$, there exists an invariant set $\Lambda \subset D$ for $\tilde{\phi}$ which is homeomorphic to S and $\tilde{\phi}$ on Λ is topologically conjugate to the shift $\boldsymbol{\sigma}$ on S.*

Note that Λ results from the multiple intersections of $\mathbf{W}^u, \mathbf{W}^s$ which lie near ∂D_0, and it is a Cantor set. This outlines the proof of Theorem 3.52. Analogous to Theorem 3.50, Xia also proved that other than the Jacobi integral, there are no other real analytic integrals for the restricted problem for all but a finite set of $\mu \in (0,1)$. See also [138] where this is also proven, for $\mu \ll 1$.

3.6.6 Existence of a Hyperbolic Invariant Set on the Weak Stability Boundary

We are interested in case (i) in Theorem 3.57, where $|C| - \sqrt{2} \gtrsim 0, \mu \gtrsim 0$. Let $\tilde{\Lambda}$ be the hyperbolic invariant set obtained for case (i).

Theorem 3.57 yields a hyperbolic invariant set $\tilde{\Lambda} \subset D_0 \subset \sum_0$. $\tilde{\Lambda}$ is the intersection of the broken homoclinic tube $\mathcal{H}$ with D_0. $\tilde{\Lambda}$ is a Cantor set of points which give rise to hyperbolically unstable parabolic orbits and lies near ∂D_0. For fixed C, μ satisfying case (i), the points of $\tilde{\Lambda}$ lie very close to P_2 at a distance r_2 beyond P_2 on the Q_1-axis. The velocity vector $\mathbf{V} = \dot{\mathbf{Q}}$ of a solution in extended phase space for $(Q_1, V) \in \tilde{\Lambda}$ need not cross the Q_1-axis perpendicularly and can cross in any direction for any given value of Q_1. When they cross perpendicularly, then $\dot{r}_2 = 0$. Let β be the crossing angle for a vector $\mathbf{V}$ on the set Λ, where $\mathbf{V}$ is projected onto the Q_1, Q_2-plane. This is the crossing angle defined in Lemma 3.42. Thus, $\beta = \pi/2$ when $\mathbf{V}$ crosses normally. For $\mu = 0$ there is a continuum of parabolic orbits crossing through a given point Q_1 with angles $\beta \in [0, 2\pi]$, $\beta \neq \pi$. By continuity of the manifold of parabolic orbits with respect to μ, the same crossing angles β will generally exist for $\mu \ll 1$.

$\tilde{\Lambda}$ therefore exists on $\{Q_2 = 0\}$ where $0 < r_2 < \delta_1$, or equivalently, $0 < |C| - \sqrt{2} < \delta_2$, $|\dot{r}_2| \geq 0$, $\mu < \delta_3$, $\delta_k \ll 1, k = 1, 2, 3$.

Set

$$I_X = \{\ r_2, C, \dot{r}_2, \mu | 0 < r_2 < \delta_1, 0 < |C| - \sqrt{2} < \delta_2,$$
$$|\dot{r}_2| \leq B_1 < \infty, 0 < \mu < \delta_3\}.$$

Note that the independent variables are given in polar coordinates by $r_2, \dot{r}_2, \theta_2, \dot{\theta}_2$, on the Jacobi integral surface for each fixed C, where on the Q_1-axis, $\theta_2 = 0$. The values of $\dot{\theta}_2$ on the set I_X are obtained from the Jacobi integral equation. B_1 is a bound on $\dot{r}_1$ which is finite since P_3 does not collide with P_2, and r_2 is bounded away from zero.

We will now prove that $\tilde{W}_H$ intersects $\tilde{\Lambda}$. The values of $\delta_k, k = 1, 2, 3$, are fixed.

Theorem 3.58 *For $e_2 - 1, r_2, \mu$ sufficiently small,*

$$\tilde{\Lambda} \cap \tilde{W}_H \neq \phi. \tag{3.72}$$

Proof. I_X is given, and therefore so are $\delta_k, k = 1, 2, 3$. The first thing we need to do is insure that the value of $|C|$ on $\tilde{W}_H$ can be adjusted to lie sufficiently near $\sqrt{2}$. This is the case if ϵ_1, ϵ_2 defined in the set I_H in section 3.6.1 are sufficiently small. We recall that

$$I_H = \{\ r_2, e_2, \dot{r}_2, \mu | 0 < r_2 < \epsilon_1, 1 < e_2 < 1 + \epsilon_2, |\dot{r}_2| \leq B_2 < \infty,$$
$$0 < \mu < \epsilon_3, 0 < \epsilon_k \ll 1, k = 1, 2, 3\}.$$

I_H is slightly modified where a bound is placed on $|\dot{r}_2|$, which is finite since on $\tilde{W}_H$ P_3 does not collide with P_2. Since ϵ_1 is a small number, then B_2 will in general be large. Because $B_k > 0, k = 1, 2$, there will exist a number $b > 0$ such that $B_k > b$, where $\dot{r}_2$ from I_X, I_H agree on the set $I_b = [-b, b]$.

To insure that $0 < |C| - \sqrt{2} < \delta_2$ in I_H, we appeal to Lemma 3.44. This implies that for each given value of ϵ_3, δ_2, we can take ϵ_1, ϵ_2 sufficiently small such that on $\tilde{W}_H$, $0 < |C| - \sqrt{2} < \delta_2$.

For each $k = 1, 2, 3$, the intervals $(0, \epsilon_k)$ and $(0, \delta_k)$ intersect in open intervals $(0, \rho_k)$, $\rho_k = \min\{\epsilon_k, \delta_k\}$. Thus, the ranges of r_2, μ from I_H, I_X intersect on the Q_1-axis where $\theta_2 = 0$. Also, $\dot{r}_2$ agree on I_b. Since $r_2, \dot{r}_2, C, \mu$ have ranges which intersect on I_H, I_X, then the commonality of values of θ_2 on I_H, I_X follows by solving for $\dot{\theta}_2$ using the Jacobi integral equation. This implies that $\tilde{W}_H, \tilde{\Lambda}$ have a nontrivial intersection. □

See Figure 3.24.

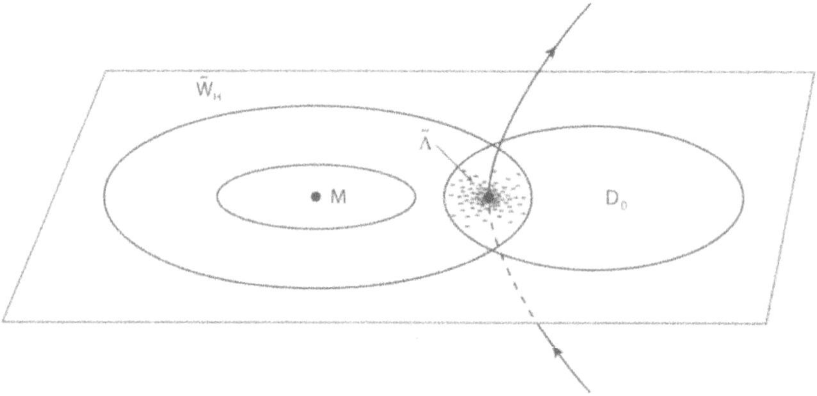

Figure 3.24 Hyperbolic invariant set $\tilde{\Lambda}$ in $\tilde{W}_H \cap D_o$. A pseudoballistic capture orbit shown passing through $\tilde{\Lambda}$ on its way to near infinity. It is lying on the invariant hyperbolic set $\tilde{\Lambda} \times \{t\}$ in the extended phase space. This is analogous to the orbit ϕ shown in Figure 3.18.

Theorem 3.58 connects the hyperbolically extended weak stability boundary with chaotic dynamics. This occurs on the component of $\tilde{W}_H$ of $\tilde{W}$ where pseudoballistic capture occurs. Thus, we have proven the following theorem.

Theorem 3.59 *$\tilde{W}_H$ intersects $\tilde{\Lambda}$ on the Q_1-axis near P_2 if $0 < r_2 < \epsilon_1, 0 < e_2 - 1 < \epsilon_2, 0 < \mu < \epsilon_3$, where $\epsilon_k, k = 1, 2, 3$ are sufficiently small. Thus pseudoballistic capture on $\tilde{W}_H \cap \tilde{\Lambda}$ implies permanent capture, unbounded oscillatory motion, and all of the motions prescribed by the sequence space S.*

The fly-bys of P_3 to P_2 when it passes through $\tilde{\Lambda}$ will all be very slightly hyperbolic since $\epsilon_2 \ll 1$. P_3 moves out close to infinity, then returns after a long time. Each time it returns to near P_2 on $\tilde{W}_H \cap \tilde{\Lambda}$, it will be pseudoballistically captured. This then yields an infinite sequence of such capture states. This is summarized in the following theorem.

Theorem 3.60 *For the orbits associated to $\tilde{\Lambda}$ which oscillate infinitely often, then each of these orbits yields an infinite sequence of pseudoballistic capture states for all time.*

Theorem 3.58 is interesting in another respect. It illustrates that the algorithm for the definition of the set $\mathcal{W}$ in subsection 3.2.2 is reflective of the transitional behavior of the map ϕ with respect to ∂D_0 defined at the end of the previous subsection. The region ∂D_0 represents a transitional location between bounded (elliptic) and unbounded (hyperbolic) motion by producing a critical value of the velocity V. Likewise, the definition of $\mathcal{W}$ produces a transition between stable motion where P_3 cycles P_2 and returns in an elliptic state, and unstable motion where it cycles and returns with a hyperbolic state or escapes P_2 and cycles P_1. The connection between $\tilde{W}$ and $\tilde{\Lambda}$ occurs on $\tilde{W}_H$. $\tilde{W}_H$ occurs near the boundary of $\mathcal{W}$ where $e_2 = 1$.

Since the set $\tilde{W}_H$ can be explicitly analytically represented, then Theorem 3.59 implies that the part of $\tilde{W}_H$ that intersects the positive Q_1-axis can be estimated. This provides a way to accurately estimate the location of hyperbolic invariant sets in an explicit fashion.

Theorem 3.59 implies that for sequences in S which give rise to motions for P_3 which oscillate infinitely often, and therefore pass by P_2 infinitely often, the crossing angles $\beta \in [0, 2\pi], \beta \neq \pi$, will vary in a random fashion due to the chaotic nature of the dynamics on $\tilde{\Lambda}$. This implies that the corresponding argument of periapsis values $\omega \in [0, 2\pi], \alpha \neq \pi$, will also vary randomly. This has an interesting possible application to cometary motion.

An application in astronomy that may be relevant is for the motions of comets far from the Sun, P_1, which move out to the Oort cloud region approximately 10,000 astronomical units from the Sun, where an astronomical unit is defined as the distance from the Earth to the Sun, about 1.496×10^8 km. The Oort cloud is hypothesized to contain billions of comets. This is so far out that a comet moving to this region from near the Sun would be approximating an ω-parabolic orbit with respect to the solar system. Modeling this as a planar restricted three-body problem, then Theorem 3.60 would say that it could in general repeatedly fall into the solar system and pass close by a given planet, P_2, say Jupiter. It would be slightly hyperbolic with respect to P_2 and pseudoballistically captured. Also, it would be approximately parabolic with respect to the Sun. Upon passage through solar periapsis, it would move out to the Oort cloud region over hundreds of years.

However, since the crossing angle $\beta \in [0, 2\pi], \beta \neq \pi$, of P_3 with the Sun–Jupiter line would vary randomly, the angle ω of the argument of periapsis at solar periapsis with respect to the Sun-Jupiter line would in general vary between $[0, 2\pi], \omega \neq \pi$, and the variation would be random. This is valid for the restricted three-body problem, and a more realistic modeling may perhaps make this effect even more pronounced. Identifying such comets as they returned would be difficult. Also, for such comets, the pseudoballistic weak capture could lead to a more stable capture where e_2 transitions from $e_2 \gtrsim 1$ to $e_2 \lesssim 1$ due to other effects such as oblateness perturbations and by modeling the gravitational perturbations of more planetary bodies.

It is remarked that although 10,000 astronomical units sounds like an immense distance, it is negligible relative to the distance to the nearest star. The nearest star system to the Sun is the alpha-proxima centauri star system which is 4.3 light years away, and one light year is 63,240 astronomical units. Thus, the Oort cloud is only approximately 4 percent of the distance to the nearest star.

Applying Theorem 3.59 to spacecraft motion may lead to trajectories for spacecraft which are permanently captured with respect to the Moon, P_2, for the restricted three-body problem. This implies that the spacecraft would orbit the Moon for all time without collision without the use of engines. Of course, the orbit would be very large and the capture orbit would be quite unstable. However a tiny maneuver performed at lunar periapsis could stablize the capture orbit and reduce its size, where at lunar periapsis one could also achieve $e_2 \lesssim 1$. Thus, one could utilize the hyperbolic invariant set on $\tilde{W}$, together with a set of tiny maneuvers to design useful ballistic capture transfers.

Many other interesting motions are possible as prescribed by sequences in the sequence space S. When the Sun is modeled, then as seen earlier in the chapter, the problem is much more complicated. The existence of the exterior ballistic capture transfer is due to the gravitational influence of the fourth body, which is the Sun. It does not go out to near infinity as a parabolic-type orbit does. However, it does go out to the weak stability boundary region of the Earth due to the Earth–Sun gravitational interaction, which can be viewed as playing an analogous role to that which infinity does for Keplerian parabolas, as a location where the periapsis distance with respect to the Sun can be increased from low Earth orbit to the radial distance of the Moon for zero ΔV. The existence of chaos for weak capture in the restricted problem sheds a little light on the sensitive capture of the exterior transfer, and on ballistic capture in general.

In conclusion, the set $\tilde{W}_H$, and more generally $\tilde{W}$, gives rise to a complicated dynamics. Numerical studies imply the existence of orbits where P_3

abruptly changes resonance types with respect to P_2 after passing through $\tilde{W}$, as well as other kinds of dynamics which are not understood [37, 29]. These motions may be related to other types of hyperbolic invariant sets not yet discovered.

Bibliography

[1] S. Abbott. *Understanding Analysis*. Springer-Verlag, New York, 2001.

[2] R. Abraham and J. Marsden. *Foundations of Mechanics*. Benjamin/Cummings, New York, 1978.

[3] R. Abraham and J. Robbin. *Transversal Mapping and Flows*. Benjamin, New York, 1967.

[4] R. Adler. To the planets on a shoestring. *Nature*, 408:510–512, November 2000.

[5] A. Albouy. Integral manifolds of the n-body problem. *Invent. Math.*, 114:463–488, 1993.

[6] V. M. Alekseev. Quasirandom dynamical systems. i, ii, iii. *Math. USSR Sbornik*, 5, 6, 7:73–128, 505–560, 1–43, 1960, 1960, 1969.

[7] V. M. Alekseev. On the possibility of capture in the three-body problem with a negative value for the total energy constant (russian). *Uspehi Mat. Nauk*, 24(1):185–186, 1969. Reviewed (in English) by J. Moser, Math. Rev., Amer. Math. Soc., 5:1071, 1970.

[8] V. M. Alexeyev and Yu. S. Osipov. Accuracy of kepler approximation for fly-by orbits near an attracting centre. *Ergod. Th. and Dynam. Sys.*, 2:263–300, 1982.

[9] D. V. Anosov and A. B. Katok. New examples in smooth ergodic theory, ergodic diffeomorphisms. *Trudy Mosk. Math. Obsc.*, 23:3–36, 1970. English translation in Trans. Mosc. Math. Soc., Amer. Math. Soc. 23:1–35, 1972.

[10] V. I. Arnold. Small denominators i. mappings of the circle onto itself. *Izvestija Akademiĭ Nauk SSSR Ser. Mat.*, 25:21–86, 1961. English translation: Translations of the American Mathematical Society (series 2), 46: 213–284, 1965.

[11] V. I. Arnold. On the classical perturbation theory and the stability problem of planetary systems. *Dokl. Akad. Nauk SSSR*, 145:481–490, 1962.

[12] V. I. Arnold. Proof of A. N. Kolmogorov's theorem on the preservation of quasiperiodic motions under small perturbations of the hamiltonian. *Russian Mathematical Surveys (Usp. Mat. Nauk SSSR)*, 18(5):9–36, 1963. (In Russian)

[13] V. I. Arnold. Instability of dynamical systems with several degrees of freedom. *Dokl. Akad. Nauk. USSR*, 156(5):342–355, 581–585, 1964.

[14] V. I. Arnold. *Ordinary Differential Equations.* MIT Press, Cambridge, MA, 1973. (Russian original, Moscow, 1971)

[15] V. I. Arnold. *Mathematical Methods of Classical Mechanics, Second Edition.* Springer-Verlag, New York, Heidelberg, Berlin, 1978, 1989. (Russian original, Moscow, 1974, Translated by K. Vogtmann and A. Weinstein)

[16] V. I. Arnold and A. Avez. *Ergodic Problems of Classical Mechanics.* W.A. Benjamin, New York, Amsterdam, 1968.

[17] S. Aubrey. *The Devil's Staircase Transformation in Incommensurate Lattices*, volume 925, pages 221–245. Spring-Verlag, Berlin, New York, 1982.

[18] S. Aubrey. The twist map, the extended Frenkel-Kontorova model, and the devil's staircase. *Physica*, 7D:240–258, 1983.

[19] S. Aubrey and P.-Y. Le Daëron. The discrete frenkel-kontorova model and its extensions. i. exact results for the ground-states. *Physica D*, 8(3):381–422, 1983.

[20] R. R. Bate, D. D. Mueller, and J. E. White. *Fundamentals of Astrodynamics.* Dover, New York, 1971.

[21] E. A. Belbruno. *Hopping in the Kuiper Belt and Significance of the* 2 : 3 *Resonance*, volume 522 of *Series C. Math. and Physical Sciences.* Kluwer, Norwell, MA.

[22] E. A. Belbruno. Two-body motion under the inverse square central force and equivalent geodesic flows. *Celestial Mechanics*, 15:465–476, 1977.

[23] E. A. Belbruno. A new regularization of the restricted three-body problem and an application. *Celestial Mechanics*, 25:398–415, 1981.

[24] E. A. Belbruno. *Regularizations and Geodesic Flows*, volume 80 of *Lecture Notes in Pure and Applied Mathematics*, pages 1–11. Marcel Dekker, New York, 1981.

[25] E. A. Belbruno. Lunar capture orbits, a method of constructing earth-moon trajectories and the lunar gas mission. In *Proceedings of AIAA/DGGLR/JSASS Inter. Elec. Propl. Conf.*, number 87-1054, May 1987.

[26] E. A. Belbruno. Examples of the nonlinear dynamics of ballistic capture and escape in the earth-moon system. In *Proceedings of the Annual AIAA Astrodynamics Conference*, number 90-2896, August 1990.

[27] E. A. Belbruno. Through the fuzzy boundary: A new route to the moon. *Planetary Report*, 7(3):8–10, May/June 1992.

[28] E. A. Belbruno. The dynamical mechanism of ballistic lunar capture transfers in the four-body problem from the perspective of invariant manifolds and Hill's regions. CRM Research Report 270, Centre de Recerca Matematica, Institute d'Estudis Catalans, Barcelona, 1994.

[29] E. A. Belbruno. *Fast Resonance Shifting as a Mechanism of Dynamic Instability Illustrated by Comets and CHE Trajectories*, volume 822 of *Annals of the New York Academy of Sciences*, pages 195–226. May 1997.

[30] E. A. Belbruno. Prediction of capture based on one parameter. Interim Report 3, Jet Propulsion Laboratory, Sept. 2000. JPL Contract No. 1213585.

[31] E. A. Belbruno. Analytic estimation of weak stability boundaries and low energy transfers. *Contemp. Math.*, V(292):17–47, 2002.

[32] E. A. Belbruno. Construction of periodic orbits in Hill's problem for $c \gtrsim 3^{\frac{4}{3}}$, 2002. to appear in New Advances in Celestial Mechanics and Hamiltonian Systems, Kluwer Academic Press, 2003.

[33] E. A. Belbruno and J. P. Carrico. Calculation of weak stability boundary ballistic lunar capture transfer trajectories. In *Proceedings of AIAA/AAS Astrodynamics Specialist Conference*, number 2000-4142, Denver, CO, August 2000. AIAA.

[34] E. A. Belbruno, R. Humble, and J. Coil. Ballistic capture lunar transfer determination for the U.S. Air Force Academy Blue Moon Mission. In *Proceedings of the AAS/AIAA Space Flight Mechanics Meeting*, volume 95 of *Advances in Astronautical Science, Spaceflight Mechanics*, Huntsville, AL, February 1997.

[35] E. A. Belbruno and J. Llibre. An estimation of the stability of planetary orbits with applications to astronomy. Unpublished manuscript, 1989.

[36] E. A. Belbruno, J. Llibre, and M. Ollé. On the families of periodic orbits which bifurcate from the circular sitnikov motions. *Celestial Mech. Dynam. Astronom.*, 60:99–129, 1994.

[37] E. A. Belbruno and B. Marsden. Resonance hopping in comets. *The Astronomical Journal*, 113(4):1433–1444, April 1997.

[38] E. A. Belbruno and J. Miller. A ballistic lunar capture trajectory for the Japanese spacecraft Hiten. Technical Report JPL-IOM 312/90.4-1731-EAB, Jet Propulsion Laboratory, June 15, 1990.

[39] E. A. Belbruno and J. Miller. Sun-perturbed earth-to-moon transfers with ballistic capture. *Journal of Guidance, Control, and Dynamics*, 16(4):770–775, July–August 1993.

[40] E. A. Belbruno, R. Ridenoure, and C. Ocampo. Historical retrospective of the hgs-1 mission. *Submitted for Publication*, 2003.

[41] E. A. Belbruno and I. P. Williams. *Asteriods and Trans-Neptunian Objects*, volume 522 of *Series C. Math. and Physical Sciences*, pages 1–4. Kluwer, 1999.

[42] M. Bello-Mora, F. Graziani, P. Teofilatto, C. Circi, M. Porfilio, and M. Hechler. A systematic analysis on weak stability boundary transfers to the moon. In *Proceedings of 51st Inter. Astronautical Congress*, number IAF-00-A.6.03, Rio de Janeiro, Brazil, October 2000.

[43] G. D. Birkhoff. Proof of poincaré's geometric theorem. *Transactions of the American Mathematical Society*, 14:14–22, 1913.

[44] G. D. Birkhoff. *Dynamical Systems*, volume 9, pages 290–291. American Mathematical Society, Providence, RI, 1927. Reprinted with an introduction by J. Moser and a preface by M. Morse, 1966.

[45] G. D. Birkhoff. Stability and the equations of dynamics. *Am. J. Math.*, 49:1–38, 1927.

[46] G. D. Birkhoff. *Collected Mathematical Papers.* American Mathematical Society, Providence, RI, 1950. Reprinted by Dover, New York, 1968.

[47] H. Bohr. *Almost Periodic Functions.* Chelsea, New York, 1947. English reprint of original in German, Fastperiodische Functionen, Springer-Verlag, Berlin, 1932.

[48] R. Bowen. Symbolic dynamics for hyperbolic flows. *American Jounal of Mathematics*, 95:429–459, 1972.

[49] A. D. Brjuno. Instability in a hamiltonian system and the distribution of asteriods. *Mat. Sbornik*, 2(125):272–312, 1970.

[50] D. Brouwer and G.M. Clemence. *Methods of Celestial Mechanics.* Academic Press, New York, London, 1961.

[51] H. Bruns. Über die integrale des vielkörper-problems. *Acta. Math.*, 11:25–96, 1887–1888.

[52] M. Capiński and E. Kopp. *Measure, Integral and Probability.* Springer-Verlag, London, 1999.

[53] J. Carr. *Applications of Center Manifold Theory.* Springer-Verlag, New York, Heidelberg, Berlin, 1981.

[54] J. Chazy. Sur certaines trajectoires du problème des n corps. *Bull. Astron.*, 35:321–389, 1918.

[55] J. Chazy. Sur l'allure finale du mouvement dans le problème des trois corps. I. Quand le temps croit indefiniment. II, III. *Annales Sci. de l'Ecole Norm. Sup. 3e Sér.*, 3(39):22–130, 1922.

[56] A. Chenciner. Orbites périodiques et ensembles de Cantor invariantes d'Aubrey-Mather au viosinage d'une bifurcation de Hopf dégénerée de difféomorphismes de r^2. *C.R. Acad. Sci. Paris I*, 297:465–467, 1983.

[57] A. Chenciner. Collisions totales, mouvements complètement paraboliques et réduction des homothéties dans le probléme des n corps. *Regular and Chaotic Dynamics*, 3(3):93–106, 1998.

[58] S. N. Chow and J. K. Hale. *Methods of Bifurcation Theory.* Springer-Verlag, New York, Heidelberg, Berlin, 1982.

[59] C. Circi and P. Teofilatto. On the dynamics of weak stability boundary lunar transfers. *Cel. Mech. Dyn. Astron.*, 79:41–72, 2001.

[60] E.A. Coddington and N. Levinson. *Theory of Ordinary Differential Equations.* McGraw-Hill, New York, 1955.

[61] C. Conley. On some new long periodic solutions of the plane restricted three-body problem. *Comm. Pure Appl. Math.*, 16:449–467, 1963.

[62] C. Conley. Low energy transit orbits in the restricted three-body problem. *SIAM J. Appl. Math.*, 16:732–746, 1968.

[63] C. Conley. *Twist Mappings, Analyticity and Periodic Solutions which Pass Close to an Unstable Periodic Solution*, pages 129–154. Benjamin, New York, 1968.

[64] C. C. Conley. On the ultimate behavior of orbits with respect to an unstable critical point I, oscillating, asymptotic and capture orbits. *J. Differential Equations*, 5:136–158, 1969.

[65] I. P. Cornfield, S. V. Fomin, and Ya. G. Sinai. *Ergodic Theory.* Springer-Verlag, New York, Heidelberg, Berlin, 1982.

[66] A. Denjoy. Sur les courbes définies par les équations différentielles à la surface du tore. *Journal de Mathématiques Pures et Appliquées*, 9(11):333–375, 1932, 1933.

[67] A. Deprit and A. Deprit-Bartolomé. Stability of the triangular lagrangian points. *Astron. J.*, 72:173–179, 1967.

[68] R. Devaney. The baker transformation and a mapping associated to the restricted three-body problem. *Commun. Math. Phys.*, 80:465–476, 1981.

[69] R. Devaney. *An Introduction to Chaotic Dynamical Systems.* Addison-Wesley, Reading, MA, 1989.

[70] R. L. Devaney. Blowing up singularities in classical mechanical systems. *Amer. Math. Monthly*, 89:535–552, 1982.

[71] F. Diacu. The slingshot effect of celestial bodies. *π in the Sky*, pages 16–17, December 2000.

[72] F. Diacu and P. Holmes. *Celestial Encounters: The Origins of Chaos and Stability.* Princeton University Press, Princeton, NJ, 1996.

[73] F. N. Diacu. Regularization of partial collisions in the n-body problem. *Differential and Integral Equations*, 5:103–136, 1992.

[74] F. N. Diacu. *Singularities of the N-Body Problem.* Les Publications CRM, Montreal, 1992.

[75] L. Dye. With a boost from JPL, Japanese lunar mission may get back on track. *Los Angeles Times, Science Section*, July 16, 1990.

[76] R. W. Easton. Generalized melnikov formulas. *J. Nonlinear Analysis Theory, Methods, and Applications*, 8(1):1–4, 1984.

[77] R. W. Easton. Parabolic orbits for the planar three-body problem. *J. Diff. Equ.*, 52(1):116–134, 1984.

[78] R. W. Easton. Capture orbits and Melnikov integrals in the planar three-body problem. *Cel. Mech. Dyn. Astr.*, 50:283–297, 1991.

[79] R. W. Easton and R. McGehee. Homoclinic phenomena for orbits doubly asymptotic to an invariant three-sphere. *Indiana Univ. Math. J.*, 28(2):211–240, 1979.

[80] L. Euler. De motu rectilineo trium corporum se mutuo attrahentium. *Novi Comm. Acad. Sci. Imp. Petrop.*, 11:144–151, 1767.

[81] R. W. Farquhar. The control and use of libration-point satellites. Technical Report TR R-346, NASA, September 1970.

[82] R. W. Farquhar, D. P. Muhonen, C. R. Newman, and H. S. Heuberger. Trajectories and orbital maneuvers for the first libration-point satellite. *J. Guid. and Control*, 3:549–554, 1980.

[83] V. Fock. *Zectschrift für Physik*, volume 98. 1935.

[84] A. Frank. Gravity's rim: Riding chaos to the moon. *Discover*, pages 74–49, September 1994.

[85] W. Gibbs. Banzai! *Scientific American*, page 22, July 1993.

[86] H. Goldstein. *Classical Mechanics*. Addison-Wesley, Reading, MA, 1950.

[87] G. Gomez, J. Llibre, R. Martinez, and C. Simó. Station keeping of libration point orbits. Technical Report 5648/83/D/JS(SC), Final Report: ESOC Contract, 1985.

[88] W. B. Gordon. A minimizing property of keplerian orbits. *American Journal of Mathematics*, 99(15):961–971, 1977.

[89] J. Guckenheimer and P. Holmes. *Nonlinear Oscillations, Dynamical Systems, and Bifurcations of Vector Fields*, volume 42 of *Applied Mathematical Sciences*. Springer-Verlag, New York, Berlin, Heidelberg, Tokyo, 1983.

[90] G. Györgyi. Kepler's equation, Fock variables, Bacry's generators and Dirac brackets. *Nuovo Cimento*, 53(A):717–736, 1968.

[91] J. K. Hale. *Ordinary Differential Equations*. Wiley, New York, 1969.

[92] J. K. Hale and H. Koçak. *Dynamics and Bifurcations*. Springer-Verlag, New York, Berlin, Heidelberg, 1991.

[93] C. T. Hall. *A Geometric Introduction to Topology*. Addison-Wesley, Reading, MA, 1972.

[94] P. Halmos. *Measure Theory*. Springer-Verlag, New York, 1950.

[95] W. R. Hamilton. *The Hodograph or a New Method of Expressing in Symbolic Language the Newtonian Law of Attraction*, volume III, pages 344–353. 1845–47. see December 1846.

[96] P. Hartman. *Ordinary Differential Equations*. Wiley, New York, 1964.

[97] M. Hénon. Numerical Exploration of the Restricted Problem. V. *Astron. & Astrophys.*, 1:223–238, 1969.

[98] T. A. Heppenheimer. Guidance and trajectory considerations in lunar mass transportation. *AIAA Journal*, 15(4):518–525, April 1977.

[99] M. Herman. Stabilitie topologique des systemes dynamiques conservatifs, a paraitre dans atas do 18°. *Col. Bras. Mat.*, 1992.

[100] M. Herman. Hamiltonian dynamics and the ergodic hypothesis. Notes from Samnel Eilenberg Lectures at Columbia University, 1996.

[101] M. R. Herman. Sur la conjugaison différentiable des difféomorphismes du cercle à des rotations. *Publications Mathématiques de l'Institut des Hautes Études Scientifiques*, 49:5–234, 1979.

[102] M. R. Herman. Sur les courbes invariantes par les difféomorphismes de l'anneau, I. *Astérisque*, 103–104, 1983.

[103] D. Hilbert and S. Cohn-Vossen. *Geometry and the Imagination.* Chelsea, New York, 1952.

[104] G. W. Hill. Researches in the lunar theory. *Am. J. Math.*, 1:5–26, 129–147, 245–260, 1878.

[105] M. Hirsch, C. Pugh, and M. Shub. *Invariant Manifolds*, volume 583. Springer-Verlag, New York, Berlin, Heidelberg, 1977.

[106] M. W. Hirsch and C. C. Pugh. *Stable Manifolds and Hyperbolic Sets*, volume 14, pages 133–163. 1970.

[107] M. W. Hirsch and S. Smale. *Differential Equations, Dynamical Systems and Linear Algebra.* Academic Press, New York, 1974.

[108] J. G. Hocking and G. S. Young. *Topology.* Addison Wesley Series in Mathematics. Addison-Wesley, Reading, MA, 1961.

[109] R. E. Hoelker and R. Silber. *The Bi-Elliptic Transfer Between Circular Coplanar Orbits*, volume 3 of *Advances in Ballistic Missles and Space Technology*, pages 2–59. Pergamon, Oxford, England, 1959.

[110] W. F. Hohmann. *Die Erreichbarkeit der Himmelskorper.* Oldenbourg, Munich, 1925.

[111] P. Holmes and J. E. Marsden. Horseshoes and arnold diffusion for hamiltonian systems on lie groups. *Indiana U. Math. J.*, 32:273–310, 1983.

[112] P. J. Holmes and J. E. Marsden. Horseshoes in perturbations of hamiltonians with two degrees of freedom. *Comm. Math. Phys.*, 82:523–544, 1982.

[113] P. J. Holmes and J. E. Marsden. Melnikov's method and arnold diffusion for perturbations of integrable Hamiltonian systems. *J. Math. Phys.*, 23(4):669–675, 1982.

[114] E. Hopf. *Ergodentheorie*, volume 5. Springer-Verlag, Berlin, New York, 1937.

[115] H. Hopf. *Selecta.* Springer-Verlag, Berlin, 1974.

[116] H. Hoschstadt. *Differential Equations: A Modern Approach.* Holt, Rinehart, Winston, New York, 1964.

[117] C. Howell, B. Barden, and M. Lo. Application of dynamical systems theory to trajectory design for a libration point mission. *Journal of Astronautical Sciences*, 45:161–178, 1997.

[118] M. H. Kaplan. *Modern Spacecraft Dynamics and Control.* Wiley, New York, 1976.

[119] A. Katok and D. Bernstein. Birkhoff periodic orbits for small perturbations of completely integrable Hamiltonian systems with convex hamiltonians. *Invent. Math.*, 88:225–241, 1987.

[120] A. Katok and B. Hasselblatt. *Introduction to the Modern Theory of Dynamical Systems.* Encyclopedia of Mathematics and Its Applications. Cambridge University Press, Cambridge, 1995. With a supplement by A. Katok and L. Mendoza.

[121] A. B. Katok. Some remarks on Birkhoff and Mather twist map theorems. *Ergodic Theory and Dynamical Systems*, 2(2):185–194, 1982.

[122] A. B. Katok. Periodic and quasi-periodic orbits for twist maps. In L. Garrido, editor, *Proceedings*, Springer-Verlag. *Sitges 1982*, Berlin, Heidelberg, New York, 1983.

[123] J. Kawaguchi. On the weak stability boundary utilization and its characteristics. In *Proceedings of 2000 AAS/AIAA: Spaceflight Mechanics Meeting*, number 00-176, January 2000.

[124] J. Kawaguchi, H. Yamakawa, T. Uesugi, and H. Matsuo. On making use of lunar and solar gravity assists in lunar a, planet b missions. *Acta. Atsr.*, 35:633–642, 1995.

[125] J. Kepler. *Astronomia Nova.* Prague, 1609. cf. Neue Astronomie, Munich-Berlin 1929, 412–413, or Gesam. werke 3, Munich, 1937, 480–482.

[126] W. S. Koon, M. Lo, J. E. Marsden, and S. Ross. Low energy transfer to the moon. *Celestial Mechanics and Dyn. Astron.*, 81:63–73, 2001.

[127] W. S. Koon, M. W. Lo, J. E. Marsden, and S. D. Ross. Heteroclinic connections between periodic orbits and resonance transitions in celestial mechanics. *Chaos*, 10(2):427–469, June 2000.

[128] E. Kreyszig. *Differential Geometry*, volume 11. University of Toronto Press, Toronto, 1959.

[129] M. Kummer. On the regularization of the Kepler problem. *Comm. Math. Phys.*, 84:133–152, 1982.

[130] P. Kustaanheimo. Spinor regularization of the Kepler motion. *Ann. Univ. Turku. Ser. AI.*, 73, 1964.

[131] P. Kustaanheimo and E. Stiefel. Perturbation theory of Kepler motion based on spinor regularization. *J. Reine Angew. Math.*, 218:204–219, 1965.

[132] J. L. Lagrange. *Oeuvres*, volume 6. Paris, 1873.

[133] A. M. Leontovitch. On the stability of lagrange's periodic solutions of the restricted three-body problem. *Dokl. Akad. Nauk. USSR*, 143:525–528, 1962. (In Russian)

[134] T. Levi-Civita. Sur la régularisation du problème des trois corps. *Acta Mathematica*, 42:99–144, 1920.

[135] A. Liapounoff. Problème général de la stabilité du mouvement. *Ann. Fac. Sci. Toulouse*, 9:203–474, 1907.

[136] J. Llibre, R. Martinez, and C. Simó. Transversality of the invariant manifolds associated to the lyapunov family of periodic orbits near $l2$ in the restricted three-body problem. *J. Diff. Equ.*, 58:104–156, 1985.

[137] J. Llibre and C. Simó. Oscillatory solutions in the planar restricted three-body problem. *Math. Ann.*, 248:153–184, 1980.

[138] J. Llibre and C. Simó. Some homoclinic phenomena in the three-body problem. *J. Diff. Equ.*, 37:444–465, 1980.

[139] L. Losco. Régularisation des collisions binaires du problème des n corps. *Celestial Mechanics*, 13:313–319, 1976.

[140] E. Macau. Exploiting unstable periodic orbits of a chaotic invariant set for spacecraft control. *Celes. Mech. Dyn. Sys.*, 2232:1–15, 2003.

[141] C. Marchal. *The Three-Body Problem.* Elsevier, Amsterdam, New York, 1990.

[142] J. E. Marsden and M. McCracken. *The Hopf Bifurcation and Its Applications.* Springer-Verlag, New York, Heidelberg, Berlin, 1976.

[143] J. Mather. Variational construction of connecting orbits. *Ann. Inst. Fourier*, 43:1349–1386, 1993.

[144] J. Mather and R. McGehee. *Solutions of the Collinear Four-Body Problem which Become Unbounded in Finite Time*, pages 573–587. Lecture Notes in Physics. Springer-Verlag, New York, Heidelberg, Berlin, 1975.

[145] J. N. Mather. A criterion for the non-existence of invariant circles. *Publ. Math. IHES*, 63:153–204, 1982.

[146] J. N. Mather. Existence of quasi-periodic orbits for twist homeomorphisms of the annulus. *Topology*, 21:457–467, 1982.

[147] C. McCord, K. R. Meyer, and Q. D. Wang. *The Integral Manifolds of the Three-Body Problem*, volume 132. Amer. Math. Soc., Providence, RI, 1998.

[148] C. K. McCord and K. R. Meyer. Integral manifolds of the restricted three-body problem. *Ergod. Th. Dyn. Sys.*, 21:885–914, 2001.

[149] C. K. McCord, K. R. Meyer, and D. Offin. Are hamiltonian flows geodesic flows? *Transactions of the American Mathematical Society*, 355:1237–1250, 2003.

[150] R. McGehee. A stable manifold theorem for degenerate fixed points with applications to celestial mechanics. *J. Differential Equations*, 14:70–88, 1973.

[151] R. McGehee. Triple collision in the collinear three-body problem. *Inventiones Mathematicae*, 27:191–227, 1974.

[152] R. McGehee. Singularities in classical celestial mechanics. In *Proc. Int. Congr. Math.*, pages 827–834, Helsinki, 1978.

[153] R. McGehee. *R. Von Zeipel's Theorem on Singularities in Celestial Mechanics Expositionae Mathematicae*, volume 4. 1986.

[154] R. McGehee and K. R. Meyer. *Twist Mappings and their Applications*, volume 44. Springer-Verlag, Berlin, 1992.

[155] R. P. McGehee. *Some Homoclinic Orbits for the Restricted Three-Body Problem.* PhD thesis, University of Wisconsin, Madison, 1969.

[156] V. Melnikov. On the stability of the center for time periodic perturbations. *Trans. Moscow Math. Soc.*, 12:1–57, 1963.

[157] W. Mendell. A gateway for human exploration of space? the weak stability boundary. *Space Policy*, 17:13–17, 2001.

[158] K. R. Meyer. Continuation of periodic solutions in three dimensions. *Physica D*, 112:310–318, 1998.

[159] K. R. Meyer. *Periodic Solutions of the N-Body Problem*, volume 1719. Springer-Verlag, Berlin, 1999.

[160] K. R. Meyer and R. Hall. *Hamiltonian Mechanics and the n-Body Problem.* Springer-Verlag, Berlin, 1992.

[161] J. Milnor. *Topology from the Differentiable Viewwpiont.* The University Press of Virginia, Charlottesville, VA, 1965.

[162] J. Milnor. Hyperbolic geometry: The first 150 years. *Bull. Amer. Math. Soc.*, 6:9–24, 1982.

[163] J. Milnor. On the geometry of the kepler problem. *American Mathematical Monthly*, 90(6):353–365, June–July 1983.

[164] A. F. Möbius. *Die Elemente der Mechanik des Himmels*, pages 28–59. Weidmannsche Buchandlung, Leipzig, 1843. Gesammelte Werke, Vol. IV, Verlag von S. Hirzel, 1887.

[165] R. Moeckel. Heteroclinic phenomena in the isosceles three-body problem. *SIAM J. Math. Anal.*, 15:857–876, 1984.

[166] R. Moeckel. Some qualitative features of the three-body problem. *Contemp. Math.*, 81:1–21, 1988.

[167] A. M. Molchanov. The resonant structure of the solar system. *Icarus*, 8:203–215, 1968.

[168] R. Montgomery. A new solution to the three-body problem. *Notices of the American Mathematical Society*, 48(5):471–481, May 2001.

[169] M. Morse and G. A. Hedlund. Symbolic dynamics. *Am. J. Math.*, 60:815–866, 1938.

[170] J. K. Moser. The analytic invariants of an area-preserving mapping near a hyperbolic fixed point. *Comm. Pure Appl. Math.*, 9:673–692, 1956.

[171] J. K. Moser. On the generalization of a theorem of lyapunov. *Comm. Pure Appl. Math.*, pages 257–271, 1958.

[172] J. K. Moser. On invariant curves of area-preserving mappings of an annulus. *Nachrichten der Akademie der Wissenschaften Göttingen, Math.-Phys. Kl. IIa*, 2(1):1–20, 1962.

[173] J. K. Moser. On the construction of almost periodic solutions for ordinary differential equations. In *Proc. Int. Conv. on Functional Analysis and Related Topics*, pages 60–67, Tokyo, 1969.

[174] J. K. Moser. Regularization of Kepler's problem and the averaging method on a manifold. *Comm. Pure Appl. Math.*, 23:609–636, 1970.

[175] J. K. Moser. *Stable and Random Motions in Dynamical Systems with Special Emphasis on Celestial Mechanics*, volume 77 of *Annals of Mathematics Studies*. Princeton University Press and University of Tokyo Press, Princeton, NJ, 1973.

[176] F. R. Moulton. *An Introduction to Celestial Mechanics*. Macmillan, London, 1914. Reprinted by Dover, New York, 1970.

[177] N. N. Nekhoroshev. An exponential estimate of the time of stability of nearly-integrable Hamiltonian systems. *Russian Mathematical Surveys*, 32:1–67, 1977.

[178] O. Neugebauer. *The Exact Sciences in Antiquity*. Brown University Press, Providence, RI, 1957.

[179] I. Newton. *Principia*. 1687.

[180] B. O'Neill. *Elementary Differential Geometry*. Academic Press, New York, London, 1966.

[181] G. K. O'Neill. The colonization of space. *Physics Today*, 27:32–40, Sept. 1974.

[182] Yu. S. Osipov. Geometrical interpretation of Kepler's problem. *Uspehi Mat. Nauk*, 27(2):161, 1972. (In Russian).

[183] Yu. S. Osipov. The kepler problem and geodesic flows in spaces of constant curvature. *Celest. Mech.*, 16:191–208, 1977.

[184] J. Oxtoby. Ergodic sets. *Bulletin of the American Mathematical Society*, 58:116–136, 1952.

[185] J. Oxtoby. *Measure and Category*. Springer-Verlag, Berlin, 1970.

[186] P. Painlevé. *Oeuvres*, volume 1. Ed. Centr. Nat. Rech. Sci., Paris, 1892.

[187] P. Painlevé. *Leçons sur la Théorie Analytique des Équations Differentielles*. Hermann, Paris, 1897.

[188] J. Palis. On Morse-Smale dynamical systems. *Topology*, 8:385–405, 1969.

[189] J. Palis and S. Smale. *Structural Stability Theorems*, volume 14, pages 223–232. Amer. Math. Soc. Publications, Providence, RI, 1970.

[190] H. Poincaré. *Les Méthodes Nouvelles de la Mécanique Céleste*, volume 1, 2, 3. Gauthiers-Villars, Paris, 1892,93,99.

[191] H. Pollard. *Celestial Mechanics*. Number 18. The Mathematical Association of America, Washington, DC, 1976.

[192] H. Pollard and D. G. Saari. Singularities of the n-body problem, I. *Archive for Rational Mechanics and Analysis*, 30:263–269, 1968.

[193] P. H. Rabinowitz. Periodic and heteroclinic orbits for periodic Hamiltonian system. *AIHP-Analyse Nonlin.*, 6:331–346, 1989.

[194] P. H. Rabinowitz. Homoclinic and heteroclinic solutions for a class of Hamiltonian systems. *Calc. Var.*, 1:1–36, 1994.

[195] C. Robinson. Homoclinic orbits and oscillation for the planar three-body problem. *J. Diff. Equ.*, 52(3):356–377, 1984.

[196] S. L. Ross. *Introduction to Ordinary Differential Equations, Third Edition.* Wiley, New York, 1966.

[197] H. Rüssman. Über invariante kurven differenzierbarer abbildungen eines kreisringes. *Nachr. Akad. Wiss. Gottingers II Math. Phys. Kl.*, pages 67–105, 1970.

[198] D. G. Saari. Improbability of collisions in newtonian gravitational systems. *Transactions of the American Mathematical Society*, 162, 168, 181:267–271, 521, 351–368, 1971, 1972, 1973.

[199] D. G. Saari. Singularities and collisions of newtonian gravitational systems. *Archive for Rational Mechanics and Analysis*, 49:311–320, 1973.

[200] D. G. Saari and N. D. Hulkower. On the manifolds of total collapse orbits and of completely parabolic orbits for the n-body problem. *J. Diff. Equations*, 41:27–43, 1981.

[201] D. G. Saari and Z. Xia. Oscillatory and super-hyperbolic solutions in newtonian system. *J. Diff. Equ.*, 82:342–355, 1988.

[202] D.G. Saari and Z. Xia. The existence of oscillatory and superhyperbolic motion in newtonian systems. *J. Diff. Equ.*, 82:342–355, 1989.

[203] K. Schwarzschild. Über die stabilität der bevegung eines durch jupiter gefangenen kometen. *Astr. Nachr.*, 141:1–8, 1896.

[204] C. L. Siegel and J. K. Moser. *Lectures on Celestial Mechanics*, volume 187. Springer-Verlag, New York, Heidelberg, Berlin, 1971. Translation by C. I. Kalme.

[205] E. Simiu. *Chaotic Transitions in Deterministic and Stochastic Dynamical Systems.* Princeton University Press, Princeton, NJ, 2002.

[206] C. Simó, G. Gomez, J. Llibre, R. Martinez, and J. Rodriguez. On the optimal station keeping control of halo orbits. *Acta Astronautica*, 15:391–397, 1987.

[207] Ya. G. Sinai. *Introduction to Ergodic Theory*. Princeton University Press, Princeton, NJ, 1977.

[208] K. A. Sitnikov. Existence of oscillating motion for the three-body problem. *Dokl. Akad. Nauk USSR*, 133(2):303–306, 1960.

[209] S. Smale. *Diffeomorphisms with Many Periodic Points*, pages 63–80. Princeton University Press, Princeton, NJ, 1965.

[210] S. Smale. Topology and mechanics, I, II, III. *Invent. Math.*, 10(4), 1970.

[211] M. Spivak. *A Comprehensive Introduction to Differential Geometry*. Publish or Perish, New York, 1975.

[212] E. M. Standish, X. X. Newhall, J. G. Williams, and W. M. Folkner. De-102–a numerically integrated ephemeris of the moon and planets spanning forty-four centuries. *Astron. and Astrophysics*, 125:150–167, 1983.

[213] S. Sternberg. *Celestial Mechanics I, II*. W.A. Benjamin, New York, Amsterdam, 1969.

[214] E. L. Stiefel and G. Scheifele. *Linear and Regular Celestial Mechanics: Perturbed Two-Body Motion, Numerical Methods, Canonical Theory*, volume 174. Springer-Verlag, New York, Berlin, Heidelberg, 1971.

[215] K. F. Sundman. Recherches sur le problème des trois corps. *Acta Societatis Scientiarum Fennicae*, 34(6), 1907.

[216] K. F. Sundman. Mémoire sur le problème des trois corps. *Acta. Math.*, 36:105–179, 1913.

[217] T. Sweetser. An estimate of the global minimum dv needed for Earth-Moon transfer. In *Proceedings of AAS/AIAA Space Flight Mechanics Meeting*, number AAS 91-101, February 1991.

[218] T. Sweetser et al. Trajectory design for a europa orbiter mission: A plethora of astrodynamic challenges. In *Proceedings of AAS/AIAA Space Flight Mechanics Meeting*, number AAS 97-174, February 1997.

[219] V. Szebehely. *Theory of Orbits*. Academic Press, New York, London, 1967.

[220] F. Takens. A c^1-counterexample to Moser's twist theorem. *Indag. Math.*, 33:379–386, 1971.

[221] K. Uesugi, J. Kawaguchi, N. Ishii, M. Shuto, H. Yamakawa, and K. Tanaka. Follow-on mission description of Hiten. In *Proceedings of the 18th International Symposium on Space Technology and Science*, pages 1723–1728, May 1992.

[222] H. von Zeipel. Sur les singularités du problème des n corps. *Arkiv för Matematik, Astronomie och Fysik*, 4(32):1–4, 1908.

[223] A. Weinstein. Symplectic manifolds and their lagrangean submanifolds. *Adv. Math.*, 3:329–349, 1971.

[224] E. T. Whittaker. *A Treatise on the Analytical Dynamics of Particles and Rigid Bodies*, volume 4th edition. Cambridge University Press, Cambridge, UK, 1959.

[225] S. Wiggins. *Global Bifurcations and Chaos.* Springer-Verlag, Berlin, 1988.

[226] A. Wintner. *The Analytic Foundations of Celestial Mechanics.* Princeton University Press, Princeton, NJ, 1947.

[227] Z. Xia. The existence of noncollision singularities in the n-body problem. *Annals of Mathematics*, 135:411–468, 1992.

[228] Z. Xia. Melnikov method and transversal homoclinic points in the restricted three-body problem. *J. Diff. Equ.*, 96:170–184, 1992.

[229] Z. Xia. A heteroclinic map and the oscillation, capture, escape motions in the three-body problem. *Adv. Ser. Nonlinear Dynamics*, 4:191–208, 1993.

[230] H. Yamakawa, J. Kawaguchi, N. Ishii, and H. Matsuo. On earth-moon transfer trajectory with gravitational capture. In *Proceedings AAS/AIAA Astrodynamics Specialists Conf.*, number AAS 93-633, August 1993.

[231] E. Zehnder. Homoclinic points near elliptic fixed points. *Comm. Pure Appl. Math.*, 26:131–182, 1973.

Index

www.ingramcontent.com/pod-product-compliance
Ingram Content Group UK Ltd.
Pitfield, Milton Keynes, MK11 3LW, UK
UKHW020649230225
455446UK00012B/113/J

9 780691 094809